21st Topical Conference on Refinery Processing 2018

Topical Conference at the 2018 AIChE Spring Meeting and 14th Global Congress on Process Safety

Orlando, Florida, USA
22 - 25 April 2018

ISBN: 978-1-5108-6427-6

Printed from e-media with permission by:

Curran Associates, Inc.
57 Morehouse Lane
Red Hook, NY 12571

Printed by Curran Associates, Inc. (2018)

For permission requests, please contact AIChE
at the address below.

AIChE
120 Wall Street, FL 23
New York, NY 10005-4020

Phone: (800) 242-4363
Fax: (203) 775-5177

www.aiche.org

Additional copies of this publication are available from:

Curran Associates, Inc.
57 Morehouse Lane
Red Hook, NY 12571 USA
Phone: 845-758-0400
Fax: 845-758-2633
Email: curran@proceedings.com
Web: www.proceedings.com

TABLE OF CONTENTS

(15a) Plant-Wide Scheduling for Profitable Emission Reduction in Petroleum Refineries 1
Jialin Xu, Jian Zhang, Sujing Wang, Qiang Xu

(15b) Using Advanced Process Control (APC) to Solve LPG Quality Issues While Optimizing the Product Distribution 2
Randy J. Cernik, Kent H. Rasmussen, Askhat M. Mergenbayev, Togzhan K. Abdeshova, Dennis L. Cima

(15c) Model Predictive Control on a Vacuum Column 3
Will McBride, Dustin Beebe

(25a) Operational Excellence in Refinery Distillation Better Decision Faster 4
Tim Shire

(25b) Three Products with High-Purity BTX Dividing Wall Columns in Operation 5
Manish Bhargava, Ashish Shah

(35a) Real Time Calculation and Closed-Loop Control of Dewpoint and Salt Deposition Temperature at a CDU, Using an Online Analyzer 11
Sebastien Osta

(35b) Data-Driven Real-Time Estimation and Optimization for Integrated Pre-Heat Exchange Trains and Distillate Cutpoints in Multiple Fractionation Columns 12
Brenno C. Menezes, Cesar S. Ikarimoto, Jeffrey D. Kelly, Galo A. C. Le Roux

(35c) Increase Profitability with Global Optimization 13
Shashi Mistry, Perry Nordh, Craig Mangan, Dian Wang, Nate Peters, John Escarcega

(35d) Decision Automation for Lubes and Asphalt Production Scheduling Using MILP with Sequence-Dependent Switchovers 34
Jeffrey D. Kelly, Brenno C. Menezes

(42a) Automation Strategies for Crude Feedstock Optimization 35
Kjell Wold, Tim Olsen

(42b) Improving Crude Oil Fouling Monitoring, Prediction and Mitigation Strategies in Refinery Preheat Trains 36
Emilio Diaz-Bejarano, Francesco Coletti, Sandro Macchietto

(42c) Improving the Crude Vacuum Heater 40
Andrew W. Sloley

(80a) Refinery Residue and Bitumen Upgrading:Gasification 41
Armen Abazajian, Andrew W. Sloley

(80b) Best Practices for Bulk and Additive Transfers in Lubricant Blending Plants 42
Meha Jha, Julie Valentine

(30c) Expanding the Role of LNG as a Shipping Fuel 43
Scott Schroeder

(96a) Using Advance Control to Monitor Plant Performance and Economics 44
Sanjay Dave, Andrew Trenchard, Christopher Webb

(96b) Process Monitoring of Batch Bioreactor Process Using Optimized Data Based Model 45
Chiranjivi Botre, Jonathan P. Raftery, Majdi Mansouri, Mohamed Nounou, Hazem Nounou, Nazmul Karim

(99a) Retrofit Strategy for Heat Exchanger Networks 46
Akpomiemie Mary Onome, Kexin Xu, Robin Smith, Lu Chen

(99b) Identifying the Causes of Acute Fouling in Refinery Preheat Trains 47
Emilio Diaz-Bejarano, Francesco Coletti, Sandro Macchietto

(99c) Improving Catalytic Reformer Heaters Â Duty Using New Technologies 51
Ashutosh Garg

(100a) Integration of Smart Manufacturing in the Advancement of Polymer Processing 52
Efstratios N. Pistikopoulos, Amit Manthanwar, Gerald S. Ogumerem, Baris Burnak

(100b) Smart Manufacturing: A Must to Remain Competitive 53
Munawar Saudagar, Minghua Ye, Sultan Al-Otaibe, Khalid Al-Jarba

(100c) Development of an Advanced Control Philosophy for Dry Gas Combined Steam Reforming, to Operate the Plant Efficiently during Feed Composition Changes 54
Jagan Mohan Rallapalli

(106a) cMIST Compact Inline Gas Treating Technology 55
Shwetha Ramkumar, Norman K. Yeh, Don P. Shatto

(106b) Novel Cryogenic CO_2 Removal- Operating Conditions and Economic Analysis 56
Farhad Fazlollahi

(106c) Dynamic Intensification of Separation Processes 57
Lingqing Yan, Thomas F. Edgar, Michael Baldea
(106d) Top Dividing Wall Columns in Process Intensification 58
Manish Bhargava, Roomi Kalita, Brad Fleming
(124a) Improving Profitability By Solving Crude and Fuel Blending Challenges 64
Meha Jha, Julie Valentine
(124b) What is the Right Gasoil Cutpoint? 65
Maria Aldescu, Al Shelton
(124c) Building Flexibility into a Refinery with a Delayed Coking Unit 76
Richard R. Heniford
(125a) Smart Systems for the Process Industries 77
Samvith Rao
(125b) Smart Damper Design Provides Better Control of Fired Heaters 78
Ashutosh Garg
(125c) Application Centralized Information Technology in Process Design Package Development 79
Minghua Ye, Munawar Saudagar, Sergio Mohedas
(137a) Advances in Utilizing Data to Improve Operations 80
Tim Olsen
(137b) Adaptive Anomaly Detection and Its Application to Online Asset Monitoring 81
Tim Butters
(137c) Dynamic Data Analysis™ of Large Scale Data to Monitor Fouling in Heat Exchanger Networks 88
Francesco Coletti, Emilio Diaz-Bejarano, Sandro Macchietto
(148a) Optimization of Refinery Preheat Trains: Predictive Maintenance and Operations Improvement 92
Francesco Coletti, Federico Lozano, Emilio Diaz-Bejarano, Sandro Macchietto
(148b) Refinery Preheat Train Fouling Management for Enhanced Energy Efficiency 95
Stephen Hall, Kumar Prashant
(148c) Optimising Productivity and Energy Consumption in Crude Oil Distillation Units with Investment-Free Solutions 96
Lu Chen, Lluvia M. Ochoa-Estopier, Leandro Labanca
(151a) Sorbent-Based Desulfurization Process for Removing Organic Sulfur Compounds 97
Gokhan Alptekin
(151b) Catalytic Reforming for Syngas Production Using Industrial CO_2: Carbon Capture and Chemical Manufacturing 98
Prashant Kumar, David Lynch, Gary Gildert, Mani Erfan
(151c) Medium Chain Alcohol Production Using Alkali Promoted Mo-Sulfide Fischer-Tropsch Catalysts 99
Tracy J. Benson, Russell R. Chianelli
(153a) How to Sell the Value of Digitalization with MPC 115
Laurie Ben, James Beall Iv
(153b) Autonomous Tuning and Firing Temperature Control for Variable Engine Operating Conditions - a New Control Scheme Designed for Optimized Engine Control in All Seasons and a Variety of Fuel Gas Compositions 116
Matthew Burton
(153c) To Cloud or Not to Cloud: That Is the Question 117
Ben Bishop, Bill Thackston III
(166a) The Journey Towards 2030, Was It to Ambitious? Update on the Impact of 2030 Vision and 2020 Transformation Program on Energy, Infrastructure and Other Projects in the Region 118
Ahmed Khogeer
(166b) Refinery-Petrochemical Integration: Organization and Culture 119
Andrew W. Sloley, Scott Schroeder
(166c) One Stone Two Birds: CO_2 to Methanol and Petrochemicals 120
Edgar Amaro Ronces
(174a) A New Era for Alkylation 121
Elizabeth Carter
(174b) Organotemplate-Free Synthesis of Hierarchical Beta Zeolites 122
Ke Zhang, Sergio Fernandez, Michele L. Ostraat
(174c) The Future Is Here: Robotic Catalyst Removal 123
Christopher R. Jansen, Andrew W. Sloley, Scott Schroeder

(178a) Using the Industrial Internet of Things to Monitor Process Equipment for Health, Performance, and Energy Consumption 124
Brian Atkinson
(178b) Using Automation Modernization As a Lever for Business Success 125
Laurie Ben
(185a) Advanced Process Modelling for Refinery-Petrochemical Integration 126
Stephen Hall, Abdullah Alghazal
(185b) Re-Thinking Benzene Control Strategies to Maximize Octane 127
Dominic Varraveto
(191a) Development of a Two-Stage Thermal Deoxygenation Process for Fuels and Chemicals from Biomass 128
M. Clayton Wheeler, Sampath Gunukula, Scott Eaton, Hemant P. Pendse, William J. Desisto, G. Peter Van Walsum
(191c) Development of Green Pathways for the Extraction of Medically Active Components Form Grape Waste 129
Kyle Cogswell, Aydin K. Sunol
(200a) Catalytic Option(s) for Hydrotreaters Struggling with Cracked-Rich Feeds 130
Steven F. Zink
(200b) Smart Equipment Integration for Improved Process Yield 131
Whyman William
(200c) Innovative Zoneflow® Reactor Technology- Gearing for Future in Steam Reforming for Hydrogen Generation 132
Sanjiv Ratan
(200d) Catalyst Characterization in the Molecular-Level Kinetic Modeling of Hydroprocessing Reactors 139
Pratyush Agarwal, Michael T. Klein
(200e) Molecular Kinetic Modeling in Hydrocracking 140
Pelin Cox
(200f) Lumped Kinetic Model for Hydrocracking of Heavy Oils over Oil-Dispersed Mo-Based Catalyst 141
Woohyun Kim, Kwang Ho Kim, Kang Seok Go, Nam Sun Nho, Sung Woo Jeong, Yong-Goo Kim
(200g) Change of Product Yield and Quality with a Variation of Residue Conversion in a Slurry-Phase Hydrocracking of Vacuum Residue 142
Kang Seok Go, Sung Woo Jeong, Yong Ku Kim, Nam Sun Nho, Woo Hyun Kim, Kwang Ho Kim
(202a) Tuning PID Controllers for Complex Integrating Process Dynamics 143
James Beall IV
(202b) Tune Your PID Loops Please! 164
Sam Dhaliwal
(202c) pH and Other Non-Linear Control Applications 165
Tom Nolan
(202d) Living on the Edge with Distributed Embedded Model Predictive Control 166
Patrick Thorpe, Shabroz Gill
(202e) Automation of Knowledge Work in Advanced Process Control 171
Alex Kalafatis, Hong Zhao, Lucas Reis
(208b) Inorganic Salts Removal from Hydraulic Fracturing Flowback Water By Cooling 172
Maoqi Feng, Kristin Silva
(208c) Metal-Hexacyanoferrate Based Nanoadsorbents for the Removal of Radioactive Cesium from Seawater 173
Hee-Man Yang, Chan Woo Park, Kune Woo Lee
(208d) A Computational Model to Simulate Performance of Industrial Clarifiers 174
Shankhadeep Das, Jen-Hsiang Kao, Bryon Barney, Mike Kidd, Mark Kuettel
(208e) CFD Modeling of Waste Water Bio-Reactors 175
Azita Ahmadzadeh, Steve Poklop, Reza Mostofi
(208f) Optimization of Water Demand Growth Fulfillment in Seawater Thermal Desalination System 176
Hassan Baaqeel, Mahmoud El-Halwagi
Author Index

Some abstracts in this book refer to images that were not made available by the publisher

512753 Plant-Wide Scheduling for Profitable Emission Reduction in Petroleum Refineries

Abstract:

Scheduling of front-end crude supply and refinery operations is an important part of petroleum supply-chain management. And in reality, emissions from refineries are significant and emission reduction is a very challenging task. It involves a large-scale complex manufacturing system and thousands of process streams. Refineries in one hand are eager to improve production solutions to leverage profitability margins in nowadays volatile market; on the other hand, they have to pursue cost-effective pollution prevention technologies to comply with stricter environmental regulations, such as air emission requirements.

Cost-effective solution strategies require emission reductions to be addressed from the entire plant point of view, where emission source generations and possible utilizations should be well balanced. But it is a very challenging task due to the large-scale complex manufacturing system, which consists of over 17 major facilities with more than 85 main operation units, and thousands of process streams. And it should focus on the comprehensively study of material, energy, and information exchanges within the entire manufacturing system, so as to identify the best solutions to the entire plant. Mendez et al. presented a simultaneous optimization approach for blending and scheduling of refinery plant.[1] The model could be either discrete or continuous formulation. Meanwhile, Pinto et al. presented a general modeling framework for petroleum supply chain optimization.[2, 3] But their work did not sufficiently characterize stream properties that are major concerns in refinery productions.

In this paper, a new methodology framework together with a new general scheduling model have been developed for emission conscious crude unloading, transferring, and processing (ECUTP) operations to achieve profitable emission-reduction scheduling (PERS) from the entire system point of view. To identify PERS strategies for refinery plants, a production scheduling model covers both front-end and refinery has to be involved. The key is to optimize the overall material and energy flows to make the plant net profit maximum, meanwhile to ensure emission source generations and utilizations to be smartly balanced. Other manufacturing constraints, such as operation specifications and inventory limits, should also be satisfied as well. Major air emissions from refineries: such as CO_2, volatile organic compounds (VOC), nitrogen oxides (NO_X), and particulate matters (PM) are characterized and quantified. Then, comprehensive analysis based on the scheduling results from the model will be conducted. Since the plant profit and emission can be quantitatively evaluated, alternative plant design and operations for PER opportunities might be raised, which may come from fundamental/theoretical analysis or industrial expertise. Based on that, the scheduling model will be modified and solved again to examine whether the proposed PER opportunity is feasible. If it is not virtually accomplished, the proposed new design or operations should be refined. Through such a model improvement iteratively, a feasible PER design or operation strategy will finally be generated.

The scheduling model is a large-scale mixed integer nonlinear programming problem (MINLP). The efficacy of the proposed emission conscious scheduling model has been demonstrated by different case studies. PERS thrusts target on emission source reductions in a profitable and systematic way, which are economically attractive, environmentally benign, and technologically viable for petroleum refineries. It may also help emission generators (not just limited to petroleum refineries) to proactively and systematically reduce their emissions to meet increasingly strict economic and environmental challenges.

References

1. Méndez, C.A., et al., *A simultaneous optimization approach for off-line blending and scheduling of oil-refinery operations.* Computers & Chemical Engineering, 2006. **30**(4): p. 614-634.
2. Pinto, J.M., M. Joly, and L.F.L. Moro, *Planning and scheduling models for refinery operations.* Computers & Chemical Engineering, 2000. **24**(9): p. 2259-2276.
3. Neiro, S.M.S. and J.M. Pinto, *A general modeling framework for the operational planning of petroleum supply chains.* Computers & Chemical Engineering, 2004. **28**(6): p. 871-896.

510708 Using Advanced Process Control (APC) to Solve LPG Quality Issues While Optimizing the Product Distribution

Abstract:

Chevron has recently completed an APC project at a Gas Plant. Not only did this provide substantial and measurable production impact to, it also provided deep insight into the Depropanizer and Debutanizer operation.

The use of well-established project execution methodology enabled this solution to be implemented in a relatively short period. The key measurements of how well this methodology was followed can be seen by the online performance and level of acceptance by operations personnel. Both measures were documented to be truly world class and could only have been achieved by the embodiment of inclusion and partnerships between the various teams, e.g. Operations, Process Engineering and Automation.

Recognizing the reality of our cost management environment, role model cost discipline was demonstrated by throughout this project which was implemented without any additional operating cost or capital expenditures. This is a demonstration of high performance behaviors at Chevron to simplify and deliver results within the cost constraint environment.

511562 Model Predictive Control on a Vacuum Column

Abstract:

In the petroleum industry, process automation plays a significant role in achieving product quality and production goals. Modern automatic control systems are comprised of multi-level hierarchy of control layers. At the lowest layer is a Distributed Control System (DCS) gathering process measurement and information, performing monitoring and PID control of process variables such as temperature, pressure and flow. Model Predictive Control (MPC) is a higher level of control, aimed at driving the plant conditions to make maximum profits by pushing the constraints of the plant while still meeting safety and environmental constraints.

The implementation of an Advanced Process Control (APC) project involves different phases namely, MPC design, pre-step testing and step-testing, model development, controller commissioning and support. Every phase is very important and it should be executed in the order specified above. The success of any APC project implementation also relies on working closely with control and process personnel of the plant and getting information from them to understand conditions and constraints involved in running the plant. In this presentation, the process description of vacuum column is presented first and the various phases of APC implementation are discussed. We will conclude with the results of the case study from this APC project on a vacuum column.

513926 Operational Excellence in Refinery Distillation Better Decision Faster

Abstract:

Operational Excellence results from making better decisions, faster, and then perfectly and consistently executing those decisions. In this paper we discuss how Simulation-based unit monitoring has been successfully applied to crude distillation units. Simulation-based monitoring enhances traditional unit performance monitoring by providing governance of data quality based on fundamental physical and chemical laws. In addition the simulation provides a richer set of analytic data which is more meaningful than raw measurements (for instance exchanger fouling factors and column performance parameters, instead of raw temperatures, pressures and flows), this allows the end users to make more insightful operational and maintenance decisions. Using this type of tool successfully requires building outputs into workflows, and we therefore discuss best practices for systematically aligning Simulation results with the decisions, tasks and work processes of relevant people in the organization.

Three Products with High Purity Dividing Wall Columns in Operation

Manish Bhargava, Roomi Kalita

GTC Technology US LLC., Houston, USA
mbhargava@gtctech.com
rkalita@gtctech.com

Dividing Wall Columns (DWC) involving less stringent product specifications have been proposed and operated in refinery operations in the past. These columns are often seen in individual columns such as naphtha splitters or reformate splitters. As the composition of product flows are flexible, these columns do not require advanced control methodology to maintain the product purity.

In recent years, DWCs are being implemented for production of high-purity aromatics such as Benzene-Toluene-Xylene (BTX) and recovery of mixed xylenes. This paper investigates the control and operation of these columns by use of a combination of plant operating data and dynamic simulation models. A grassroots middle dividing wall column is used in both of the cases under consideration. The aim is to identify the most favorable and economic method of operating such a column for the maximum profits and efficiency. The advantages of DWCs are evident. In a majority of the cases, DWCs have proven to generate 20-30% energy and capital investment savings as compared to conventional techniques. However, high purity aromatics extraction adds a new complexity to the operation of dividing wall column.

Introduction

Aromatics production units usually consist of a two-column sequence. For a mixed xylenes recovery unit, toluene is separated in the first column followed by a xylene column. Mixed xylenes and the C_{9+} cut are removed in this column. For a BTX separation, the bottoms from the extractive distillation unit are processed in a benzene column followed by a toluene column. High purity benzene, toluene and xylenes are produced in that sequence.

These high purity separations can be done in a middle dividing wall column, which has a vertical wall running through the central part of the column. Essentially two sections are created within the column. These can be symmetrical or asymmetrical based on the degree of separation required and the feed quality. The pre-fractionation afforded by the wall on the feed side helps in better separation and component recovery. The wall forces the middle boiling cut (either toluene or mixed xylenes) to concentrate on the other side of the wall with minimal adulteration from the lightest and heaviest components.

1.1 Mixed xylenes recovery unit installed at Tonen General, Japan

The grassroots GT-DWC® unit separates a mixed xylenes stream from the heavy reformate (Bhargava et al., 2016). Toluene and the C_9+ cut are the top and bottoms products. The DWC column pressure is elevated so that the overhead vapors of the column are used to heat the reboilers of the upstream columns. This presents an extra challenge to the operation of the DWC. Besides maintaining the high purity product specifications, the refiner also needs to generate sufficient vapors for the stable operation of the two upstream columns.

The TonenGeneral Group has an operational middle dividing column that recovers mixed xylenes from a full-range reformate feed (Figure 1). The new column is downstream of a pre-existing benzene extraction unit (BEU). The unit has two columns that recover light reformate, benzene and heavy reformate. In addition, the two existing columns operated on steam reboilers. The refinery, being located in Japan, had plot space constraints. Hence, the conventional two-column solution was abandoned in favor of a compact DWC.

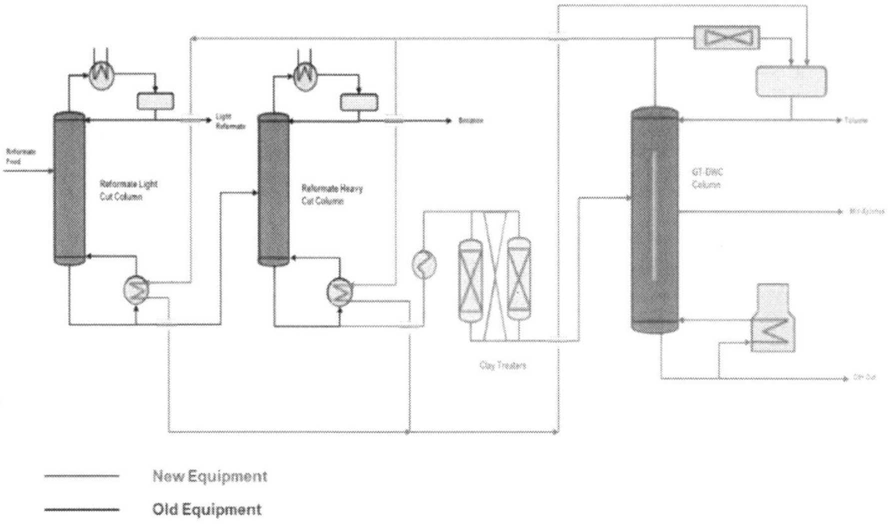

Figure 1: New GT-DWC® mixed xylenes column was installed downstream of the existing light and heavy cut columns in the Benzene Extraction Unit (BEU)

In spite of the complexities, the operation and control of a dividing wall column is fairly simple. The control philosophy for the column is shown in Figure 2.

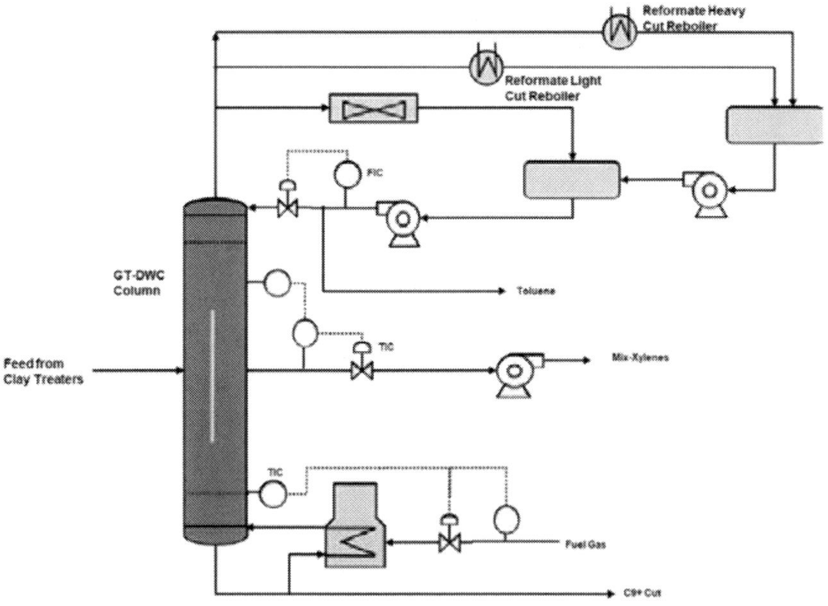

Figure 2: Control Scheme of the GT-DWC® mixed xylenes column

Since the DWC is a three-cut column, the column is defined by three primary control loops. These include the following:

1. The column operates on a fixed reflux rate. Fixing the reflux rate helps to generate consistent vapors for the upstream reboilers and thereby, the product quality of the two columns remains stable.

2. Since the column requires a fixed reflux rate, the heat duty input of the DWC column needs to be defined. This is done by cascading the fuel gas flow of the fired heater to the column bottoms temperature.

3. The column produces high purity mixed xylenes. This is achieved by cascading the side-cut product rate to one of the upper tray temperatures.

Figure 3a: High Purity Mixed Xylene Column (Operational since March 2016);
Figure 3b: Off-center Dividing Wall

The DWC is equipped with a combination of proprietary internal and external liquid splitting devices. The column maintains operational flexibility for drastic change in feed composition or product specs by utilizing this extra degree of control in such scenarios.

Figure 4 below provides the test run data of mixed xylenes product. C_8 aromatics product purity containing ethylbenzene is maintained at around 99 wt%. The recovery rate of C_8 aromatics at plant start-up is around 80 to 90 wt%. After operational adjustments and controller optimization, TonenGeneral currently maintains a high recovery rate of 95-99 wt%.

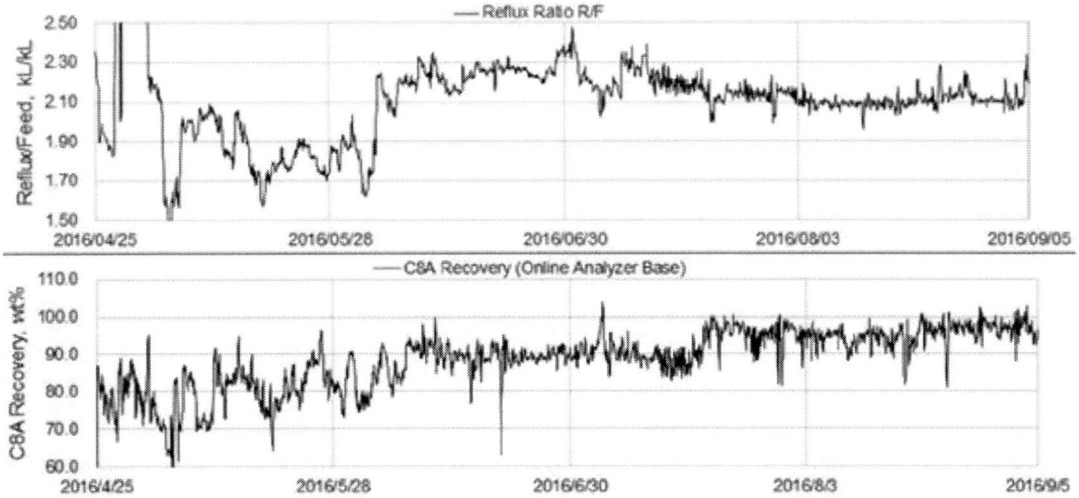

Figure 4: Plant operating data of the GT-DWC® mixed xylenes column

Since its start-up, the DWC unit maintained a consistent product specification as shown in Table 1.

Table 1: Product Specification of Mixed xylenes

Components	Units	Specifications	Remarks

C_8 aromatics	wt%	98.5	Minimum
Ethyl Benzene	wt%	18.0	Maximum
Non-aromatics	wt%	1.0	Maximum
C_9 aromatics	wt%	0.3	Maximum

1.2

1.3 Benzene-Toluene-Xylene separation unit

A grassroots middle dividing column producing high purity benzene, toluene and mixed xylenes is under construction in China for Dongying Yatong Petrochemical Co. Ltd. The startup of this column is planned for March 2018.

The column will recover benzene and toluene products in a single column from an aromatics rich extract. For operational flexibility, the column uses a combination of external and external liquid splitting. The three products required in this particular column has precise specifications and hence, the column controls need to be very specific to regulate the products.

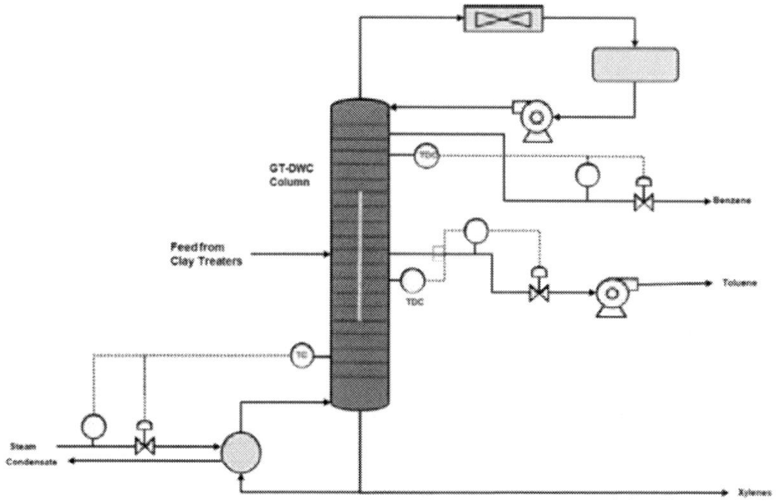

Figure 5: Control Scheme of the GT-DWC® BTX column

The column will utilize a simplified control scheme to produce on spec products as shown in Figure 5.

First, the column produces high purity benzene as the top product. To maintain the strict product specifications, the benzene product rate is cascaded to the differential temperature on the upper trays of the column.

Next, the toluene product flow is cascaded to the differential temperature of trays below the side cut on the product side of dividing wall.

Last, the steam flow rate to thermosyphon reboiler is cascaded to the column bottoms temperature.

To validate the control scheme, a dynamic simulation model has been developed in Hysys. The simulation model predicts the performance of the column based on several factors including column top pressure, reflux drum level, column bottom temperature, column sump level and the bottoms product flow. The feed flow and composition were varied to observe the shift in column behavior. The simulation model has been adjusted to match the steady state simulation results (Table 3) as closely as possible.

The model simulations are based on a four-column system to mimic the operation of the different sections in a middle DWC. Additionally, conservative estimations have been made to account for the pump characteristics and control valve pressure drops.

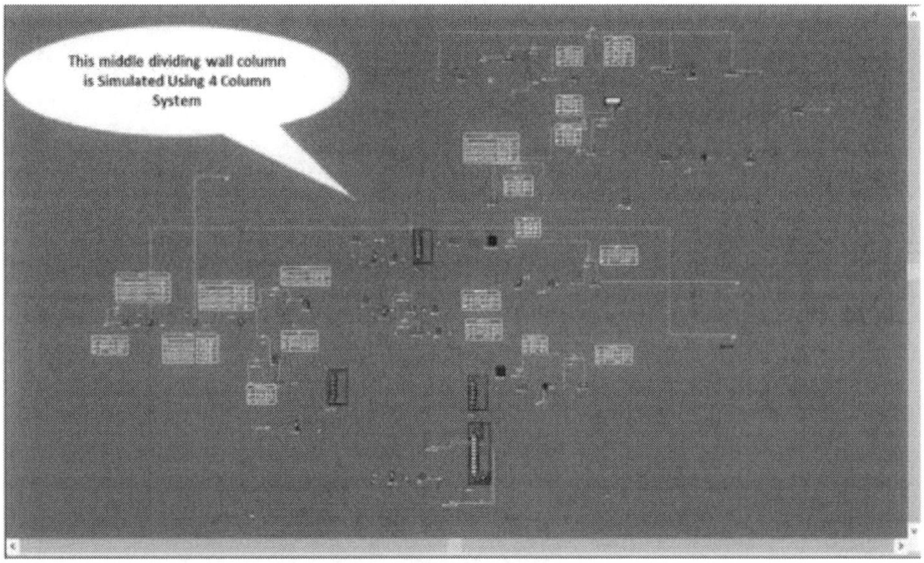

Figure 6: Four-Column Dynamic Simulation Model of BTX DWC Column in Hysys

The control scheme was validated to achieve the product specifications outlined in Table 2.

Table 2: Product Specification of GT-DWC BTX Column

Components	Units	Specifications	Remarks
Benzene Product	wt%	99.95	Minimum
Toluene Product	wt%	99.920	Minimum
Xylene Product (IBP/FBP)	°C /°C	137.5 – 141.5	

Dynamic simulation model studied the behaviour of column controls by varying the feed rate from -10% to10%. The results are presented in Figure 6a and 6b.

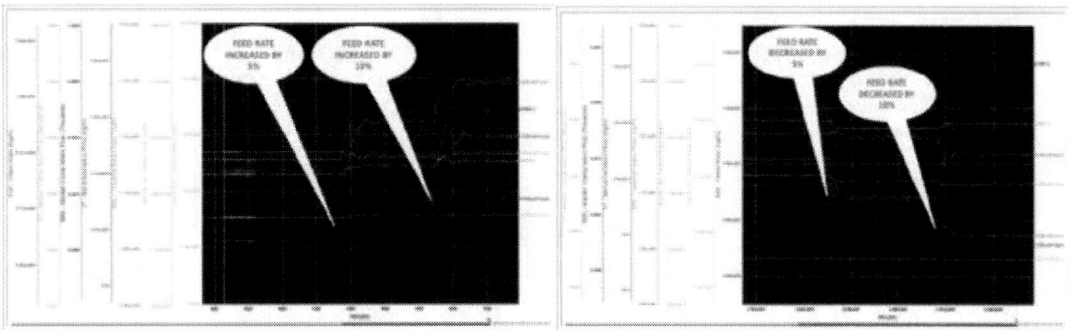

Figure 7a: Dynamic simulation response for step change in feed increase of 5% and 10%;
Figure 7b: Dynamic simulation response for step change in feed decrease of 5% and 10%

The dynamic simulation also validated the control scheme behaviour for step change in feed compositions. The results are presented in Figure 8.

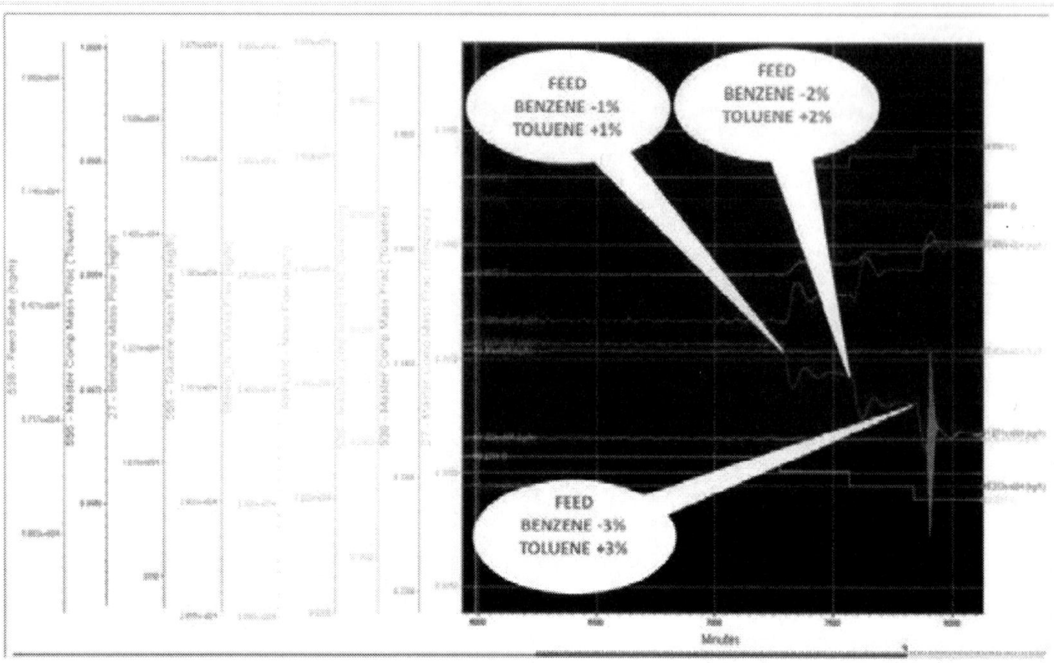

Figure 8: Dynamic simulation response for step change in benzene and toluene composition in Feed.

The dynamic model corroborates the design control scheme for the DWC column. The column generates high purity benzene and toluene products even with a +/- 10% variation in the design flow. The product purities are maintained even when the benzene/toluene composition in the feed varies by about +/-3%. The dynamic response for all the control loops evaluated is normal and as per expectations of the design.

This paper will further provide the comparison of operating data with simulation results after the start-up due in March 2018.

Conclusions

Much research has been done on Dividing Wall Columns till date in conjunction with successful implementation in several refinery applications. Despite this, a lot of refiners are apprehensive of using these innovative columns in high-purity separations. Contrary to prevalent belief, DWCs are similar in operation to regular distillation columns. The product quality in these columns can be maintained by using a simple three-control methodology. Plant operating data and simulation model results endorse the use of these controls.

References

Bhargava M., Kalita R., Gentry J.C., Suzuki N., 2016, Improved Distillation Efficiency, PTQ (October)
Bhargava M., Kalita R, Gentry J.C., 2017, Dividing Wall Column Provides Sizable Energy Savings, Chemical Processing (May)

510571 Real Time Calculation and Closed-Loop Control of Dewpoint and Salt Deposition Temperature at a CDU, Using an Online Analyzer

Abstract:

Corrosion of a CDU column overhead system can have dramatic consequences for refinery operation. Presence of water, chlorides, ammonium, amines, etcâ €¦ creates a corrosive environment requiring overhead system monitoring. Since there is a significant economic penalty associated with increasing the overhead temperature to minimize the deposition of salts, refiners are typically trying to limit this temperature increase to control the presence of some key chemical species, and measure their remaining concentrations. This presentation addresses some of the challenges associated with computing water dewpoint and ammonium chloride salt deposition temperature in real time, and shows how TOTAL has successfully achieved closed-loop control of these temperatures using a NALCO online chlorides and ammonium analyzer coupled to an online calculation software and a multivariable predictive controller.

513585 Data-Driven Real-Time Estimation and Optimization for Integrated Pre-Heat Exchange Trains and Distillate Cutpoints in Multiple Fractionation Columns

Abstract:

In crude-oil refineries, the fine adjustment of the operational variables of the hydrocarbon fractionation columns can be optimized in an integrated multi-unit on-line strategy. It considers demands on amounts and properties of intermediate hydrocarbon streams for a) downstream process-shops (e.g., to match quality feed needs in hydrotreating units) and b) blend-shops of final streams as fuels, lubes, asphalts and petrochemical feeds. However, real-time optimization (RTO) strategies in the hydrocarbon refining industry are normally modeled using first principle formulas considering reduced scale problems as for a sole process unit. Nevertheless, due to the convergence difficulties of the solutions, the RTO applications in industry are often found in off-line mode. The lack of high quality techniques of both estimation (data reconciliation, parameter estimation) and statistical analysis (redundancy, observability, variance) is the main reason for that, although difficulties from nonlinearities of rigorous formulation may arise. To enlarge the scale of the RTO from one process unit to a plant or more integrated units, this work uses actual process or operating data with reduced or relaxed-ordered models (ROM) to optimize in real-time initial and final cutpoints [1] of distillates of multiple fractionation columns integrated with their pre-heat exchange trains.

In the integrated data-driven RTO strategy proposed, the setpoints of material flows, initial and final boiling points of distillates (cutpoints), parallel split ratios of pre-heat exchanger trains, feed and tray temperatures in the hydrocarbon fractionator furnace and tower, are determined considering on-line measurement feedback to incrementally situate the plant/sub-plant to a more profitable operating or processing space. The RTO formulation considers nonlinear programming (NLP) to solve on-line (real-time, closed-loop) and in-line (real-time, open-loop) applications. By adjusting pressure, temperature, flow, separation efficiencies, catalyst activity, etc., slight ranges of separation and conversion of the hydrocarbon molecules can be achieved. Therefore, the integrated RTO manipulates operational variations in physical and chemical processes for production of crude-oil refined streams containing a range of light to heavy hydrocarbon molecules whose extension yields the amounts and properties of bulk fuels such as gasoline, kerosene, jet and diesel [2].

The proposed multi-unit data-driven RTO uses the surrogate approach of distillation blending and temperature cutpoint formulation from Kelly et al. [1] conjugated with rigorous modeling of pre-heat exchanger networks, resulting in a hybrid RTO model as found in Mahalec et al. [3], although their formulation is only applied for a single distillation unit. The surrogate or reduced order model using distillation curve adjustment or shifting technique is based in experimental ASTM / SIMDIST points and measurement of material flows of the column distillates. This is integrated with the heat exchange RTO that uses the real-time estimation (RTE) engine to generate UA heat transfer coefficients continuously and simultaneously with the data reconciliation steps.

The use of RTO is challenging and expensive because of the difficulties in building and adapting accurate models for complex chemical processes [4] and since it involves steady state detection, state/parameter estimation, data reconciliation and solving of nonlinear optimization problem on-line [5].

[1] Kelly, J. D.; Menezes, B. C., Grossmann, I. E. Distillation blending and cutpoint optimization using monotonic interpolation. *Industrial & Engineering Chemistry Research*, 53, 14146–14156, 2013.

[2] Menezes, B. C.; Grossmann, I. E.; Kelly, J. D. Enterprise-wide optimization for operations of crude-oil refineries: closing the procurement and scheduling gap. *Computer Aided Chemical Engineering*, 40, 1249–1254, 2017.

[3] Mahalec, V.; Sanchez, Y. Inferential Monitoring and Optimization of Crude Separation Units via Hybrid Models. *Computers & Chemical Engineering*, 45, 15–26, 2012.

[4] Chachuat, B.; Srinivasan, B.; Bonvin, D. Adaptation strategies for real-time Optimization. *Computers & Chemical Engineering*, 33 (10), 1557–1567, 2009.

[5] White, D. C. Online optimization: What, where and estimating ROI. *Hydrocarbon Processing*, 76 (6), 4351, 1997.

Refinery Real Time - Product Value - Global Optimization

Shashi Mistry, Craig Mangan, Nate Peters, Dian Wang

Suncor Energy, Edmonton, Canada

Perry Nordh, John Escarcega

Honeywell Process Solutions, Calgary, Canada, Phoenix AZ

Introduction

The refinery optimization system has become a key layer of controls for refinery wide product value optimization (PVO). Depending on the technology used, the optimization layer provides seamless connectivity between business and logistics information to the plant wide basic and advanced controls systems that must include both product and feed blending systems. When such system is integrated with refinery wide steady state and dynamic constraints, crude and product blending specifications, and economic data, the global product value optimization can be achieved safely and reliably. In today's technology driven world, a "smart refinery" can be then realized with this essential layer of optimization that is integrated with smart technologies such as soft sensors, multi stream Near Infrared (NIR) and Nuclear Magnetic Resonance (NMR) analyzers, wireless information from critical equipment, short and long term process predictive models data coupled with artificial intelligence (AI). The global optimization may include utilization of short, medium and long term information for end-to-end refinery asset optimization that performs not just day to day optimization but between successive refinery turn-arounds. Process models for fixed bed semi-regenerative processes such as hydrocrackers, reformers, and hydro-treaters are deployed with long term predictions of non-linear yields and catalyst deactivation allowing the refinery to operate with optimum catalyst activity strategies between turn arounds.

Suncor Edmonton Refinery, which has been involved with such optimization system over the last few years, is the subject of this paper.

Overview

Suncor Energy Inc. Edmonton refinery operates 142,000 Barrels per day (22,200m3/day) mix of light sweet and heavy sour crude from Fort McMurray oil-sands. A simplified schematic is provided in figure-1 showing process unit configuration. The products from the refinery are supplied throughout Western Canada using rail, roads and pipe-lines.

The sour crude unit is designed to process blends of Suncor synthetic and heavy sour crudes. The sweet crude unit processes treated synthetic crude that is a mix of Naphtha, Distillates and Gas oil from Oil Sands near Fort McMurray. The products include the full range of gasoline, diesel, jet fuels and light ends produced for year round specifications. The refinery operates with minimum product inventory configurations.

Agility and Optimization Technology

Agility is a primary objective of any technology used to improve optimization and reduce inventory, however in many industries this is a multidimensional and intricate problem.

INCREASE PROFITABILITY WITH GLOBAL OPTIMIZATION

Practitioners and researchers have discussed and summarized issues in unit and multi-unit optimization with various innovations over the years bringing progress to this important issue[3,4,5].

Some of the originally defined issues include:

Feed properties and characterization

Availability of good models

Accurate first principles models

Data reconciliation

Steady state detection

Intermediate pricing

Prediction of steady state values

Feasible move plan to the ultimate optimal steady state

Prediction of process responses to planned changes

Local versus global optimization

Availability of people

Progress has been made on many of these issues, including the incorporation of rigorous process models in the calculation of partial derivatives to match the controller gains to the current operating point of the process[6]. Thus, many of the issues highlighted here have been dealt with in various ways by multiple vendors; however the fundamental problem of producing the right volume/mass, at the right time, at the right quality still remains a challenge for many manufacturers.

The Optimal FEASIBLE Solution

If we define a "good" solution to be one that:

Delivers an optimal result that is feasible (respects the laws of physics and real plant limitations)

Keeps focus on the end product volume/mass, product quality, and timing of when the product is required for shipment/transfer

Simplifies the overall problem to ease the burden of understanding and maintenance over time

What if the solution could incorporate existing control layers (since these layers already deliver a feasible solution) and incorporate the end product goals (volume/mass, quality, time) - adding information from the nonlinear simulation models where possible?

A layered solution is necessary.

Optimization at Suncor Edmonton refinery

The wide range of petroleum products and specifications coupled with global market conditions presents both challenges and opportunities within the production facility and the supply chain. Creating a value chain in a production/operating facility become critically important and online optimization fits well in this important chain.

Before refinery operations can be fully optimized, it is essential that all product blending operations are dynamically optimized (see Figure 3 thru Figure 5). A refinery must be able to utilize and blend all products without downgrading or giving away valuable products. Once the dynamic blending system optimization is in place, the upstream process of integrating the process units with all their component rundowns can take place to minimize component inventory and giveaways producing just product the blender is demanding. Figure-5 shows multi header inline/online rundown diesel blending controllers deployed at Edmonton refinery without any component tanks that is linked to global online optimization.

INCREASE PROFITABILITY WITH GLOBAL OPTIMIZATION

The Online Refinery optimization technology, whether dynamic or steady state based, depends upon numerous factors including the refinery configuration, level of basic instrumentation, process complexities and operating flexibility. In all cases, global optimization benefits can only be sustained when not compromised by technology selections, software maintenance and complexities, optimization down time due to process model complexities and interactions. It is important to note that the combination of the selected optimization system and all underlying layer of Advanced Process Controller (APC) inter-operability are required to be maintained without degradation of the performance and controllability. This is especially true when multiple process units are required to be integrated for global optimization system. With this discussion in mind, we can define optimization benefit achievable within the refinery operating boundary as follows:

Realizable Optimization Benefit = (Potential benefit− Current benefit) x Service Factor (SF)

It can be seen that the actual realizable benefit depends on a selected technology and its inherent limitations. The overall performance or benefits are ultimately governed by combination of service factors (SF) that may be defined as follows:

$SF = f(MQ, GE, SD, PE, ICD, SPD, RTOT \ldots.)$

Where: SF – Overall maximum service factor of the selected technology achievable

MQ – Process Model quality service factor

GE – Gross Error Detection service factor

SD – Steady State Detection service factor

PE – Parameter Estimation / Identification service factor

ICD – Inter Controller Dependency service factor

SPD - Ability to execute set-point within real-time optimization (RTO) that is consistent with MPC service factor

RTOS - Stability of MPC controller and RTO service factor

The ability to optimize a large process operating boundary are very critically governed by some or all of these service factors. Some of the issues related to service factors and the traditional approaches to optimization are needed to be revisited and redefined. Also, as the optimization scope grows within the global operating boundary, the number of operating limits and constraints can grow exponentially not to mention their complexity. Therefore, any technology applied to the optimization must be manageable, maintainable and fit for intended use. Traditional steady state RTO may not meet these challenges so easily. Gate to gate optimization necessarily requires bringing the planning, controls and optimization activities closer together, creating a new platform of online production and controls that satisfies customer demand with greatest agility.

It is evident that combination of the service factors listed above is technology specific and can potentially compromise the intended use of the optimization. Cycle time all-or-nothing solution associated with some of these RTO technologies is generally not acceptable to refinery operations from safety and reliability prospective. The optimization solution must be robust, repeatable, predictable, implementable, and seamless that adds values to refinery bottom line. The refinery optimization requires both fast and slow dynamics to be equally incorporated and must be adequately handled by the system. This is often overlooked by traditional RTO system and when included, may not provide robust solution. When large numbers of processes are required to be integrated to a traditional optimization system, the issues of global data reconciliations, multi-process steady state, multi-process gross error detection, intermediate inventories, integrating variables often becomes unsolvable and/or unusable (see figure-6). The process disturbances and process lag between multiple processes linked together can become very complex and the actual steady state solution often difficult to attain and implement. Cat and mouse game between steady state optimization solutions and lower tier dynamic controllers however independently robust must be avoided for safe and reliable operations of the plant. Also, no process model is perfect throughout its entire operating trajectory (start-of-run to end-of-run time scale). Frequent tuning and maintenance makes them un-usable especially when directly coupled to optimization tools. A robust model that accurately describes process from start of run to the end-of-run with process gains generally provides stable solution even though de-coupled from optimization and MPC controllers. Figure 7 shows a semi-regenerative process model providing short and long term constraints required for optimization which are often forgotten in the traditional optimization design.

Highlights of Technology Used

Suncor Edmonton Refinery deploys Honeywell's Profit Optimizer Technology (Dynamic Quadratic Program (DQP)) for global optimization of the refinery. Figure 8- thru figure-11 shows how refinery wide APC Controllers are integrated to the DQP environment during the initial prototype and testing phases. Figure-9 shows how multiple controllers are linked inside DQP using bridge modeling (BM) and variable cloning methodology with consistent process dynamics. Figure-10 shows schematically how non-linear process gains are mapped to MPC controller for all fast/slow dynamics. This methodology is used for all blenders as well as processes with highly nonlinear process behaviors at varying dynamics. Note that these models are decoupled or independent from the optimization solver or the APC controller system runtime environment. This provides robust integration of non-linearity of individual processes at advanced controller (APC) level and do not require complex treatment that steady state optimization necessitates. The segregation of the gain mapping (see figure 11) for each process area and the APC controller from the optimization layer tools provides modularity that adds stability to the system as the process model can be independently worked on without compromising the runtime or the stability of either the system or the optimization overall trajectory in operations. It is important to note that the design of the entire controllers and optimization system can be changed with relative ease with such design. The operator and engineer interface for optimization system need to have the same look and feel of the APC controllers allowing ease of understanding, operating and monitoring by all.

Profit Optimizer

When solving multi-unit and plant-wide optimization on a minute-by-minute basis, it is critical to rely on measured process relationships and fully leverage dynamic process information that resides in the underlying Profit Controller applications. Using this approach, economic benefits of traditional RTO can be delivered with certainty of feasibility based on the optimization feedback from the underlying controllers.

The optimization can be set up to communicate between layers in 3 separate ways:

1. Cooperative Connection (default):
    a. When to use: The optimal steady state solution may not be feasible within underlying controller's time horizon
2. Direct Connection:
    a. When to use: Optimal solution always feasible or transient dynamic constraint violation is okay

3. User Connection:

    a. When to use: to force a local optimization that is known to be correct to lead the overall optimized solution. E.g., maximize product rates on most downstream unit to more quickly relieve high capacity constraints upstream

Project Details and results

Since the Edmonton Refinery does not have rundown components tanks, it was necessary to have all blending rundown controllers and non-linear gain mapping applications integrated to global optimization DQP system during the initial phase of the project. This allowed additional benefits to be realized as product pool requirements for gasoline, diesel and jet fuels could be immediately targeted by DQP with built in PVO drivers, with appropriate tradeoffs in real time. This approach allowed just-in-time productions of all required blends to be achieved [1, 2] (see reference-1 for details). Benefits associated with such optimization include minimum or no need of component tanks and equipment, consistent operational utilization of all products, rapid and frequent changes to target specifications. An example being a large amount of RVP target changes implemented with optimum productions rates for multiple gasoline headers. Here different RVP specifications with minimum giveaway and within required product specifications are achieved (see figure-13) direct from the production. This optimization configuration avoids component tanks usage and delivers economic benefits with lowest specifications give away that are often not associated with component tanks. Figure-16 and Figure-17 shows DQP optimizing C5C6 component to gasoline inline blend pool based on the product price shift, keeping all other refinery wide variables in mind. With component tanks, there is often a disconnect between blending systems and production units, as tanks serve as buffer that may result in individual process optimized at odds with blending requirements (suboptimal global situation). The profit optimization (DQP) integrated with plant advanced process controllers (APC) having the same process models and dynamics work in tandem, to optimize individual processes towards global productions targets, by manipulating rundown components flows in coordination with plant productions with economic drivers and product specification. The ultimate goal is to produce a precise volume, at a precise quality, at a precise time to deliver a specific product to market – requiring minimum tanks and less production time/energy within the facility whist at the same time minimizing quality giveaways.

After an initial optimization success with implementation of entire inline blending system, Suncor continued to add upstream process units to the system starting with the gasoline producing processes of the refinery. With all other remaining process units integrated to the system will provide Suncor Energy Inc., with one of the largest online optimization network in place. The automated closed loop optimization will be continued beyond the crude unit(s) once the on-line crude analysis project is completed toward the year end. The crude feed selection and blending controls will then provide end-to-end optimization at a specific quality at a specific time with least cost feedstock at all times.

Conclusion

Edmonton Refinery real-time global optimization has come a long way and while challenging, the economic benefits offered in the initial phase are substantial. The DQP based optimization allows Suncor Energy Inc. Edmonton Refinery to bridge the gap between planning and productions in real-time in constantly changing operating and economic conditions with stable controls to realize the product value optimization (PVO). Edmonton Refinery Vice President re-iterated as follows:

INCREASE PROFITABILITY WITH GLOBAL OPTIMIZATION

"The current economic environment requires all energy producers to make use of the latest innovative technology to maximize products and stay competitive. What used to be done to optimize and drive the refinery will not be competitive in tomorrow's refinery. At the Edmonton Refinery we have taken significant steps forward over the last 6 years to move towards true on-line optimization and implementing a multi-unit application to drive the product blending is the latest milestone. We will continue to move forward with the end goal of a refinery wide real time optimization application as a cornerstone of our "smart refinery" configuration."

Dean Wilcox,

Vice President, Suncor Edmonton Refinery

The DQP based optimization has empowered understanding among working groups dealing with day to day optimization as well as long term strategy with eventual goals of significantly improved bottom line.

Acknowledgement

The authors would like to thanks Suncor Energy Inc. and Honeywell management for permission to publish this work. In addition, we would like to thank Automation, Planning and Production group at Suncor Edmonton refinery for their invaluable contribution to the success of the project.

References

[1] Mistry Shashi, Mangan Craig, Eshpeter Jim (Suncor Energy Inc., Sharma Sanjay (Honeywell) *"Real-time optimization of Gasoline run-down headers blending operation" AICHE Spring 2013*

[2] Mistry Shashi (2010-12) IPCC Inc., *"Gasoline Blending Technology development for On-line Applications"*

[3] Friedman, Y. Z., *"What's wrong with closed loop optimization?" Hydrocarbon Processing, October 1995*

[4] Friedman, Y. Z., *"Closed Loop Optimization Update" Petrocontrol editorial, 2000*

[5] Lu, Joseph, *"Closing the Gap between Planning and Control: a Multiscale MPC Cascade Approach, IFAC World Congress, Cape Town, August 2014*

[6] Nath, N., Alzein, Z., Pouwer, R., Lesieur, M., *"On-line Dynamic Optimization of an Ethylene [6]Plant Using Profit Optimizer", NPRA Computer Conference, November 1999.*

Figure-1 shows the simplified configurations of the refinery from crude to blending.

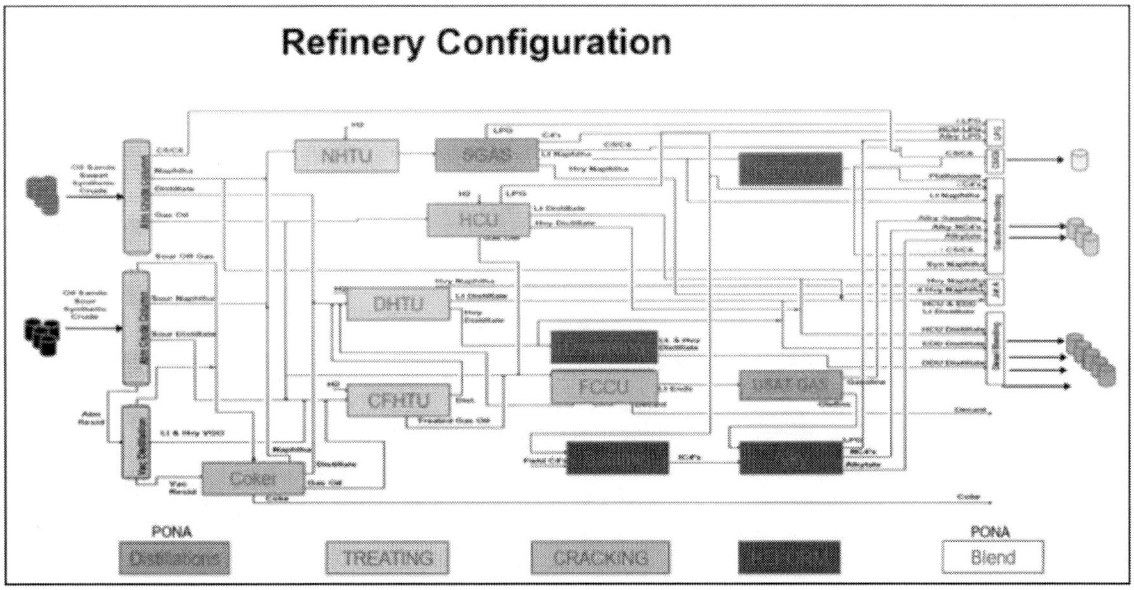

Figure-2 Steady State Issues related to single and multiple process connections

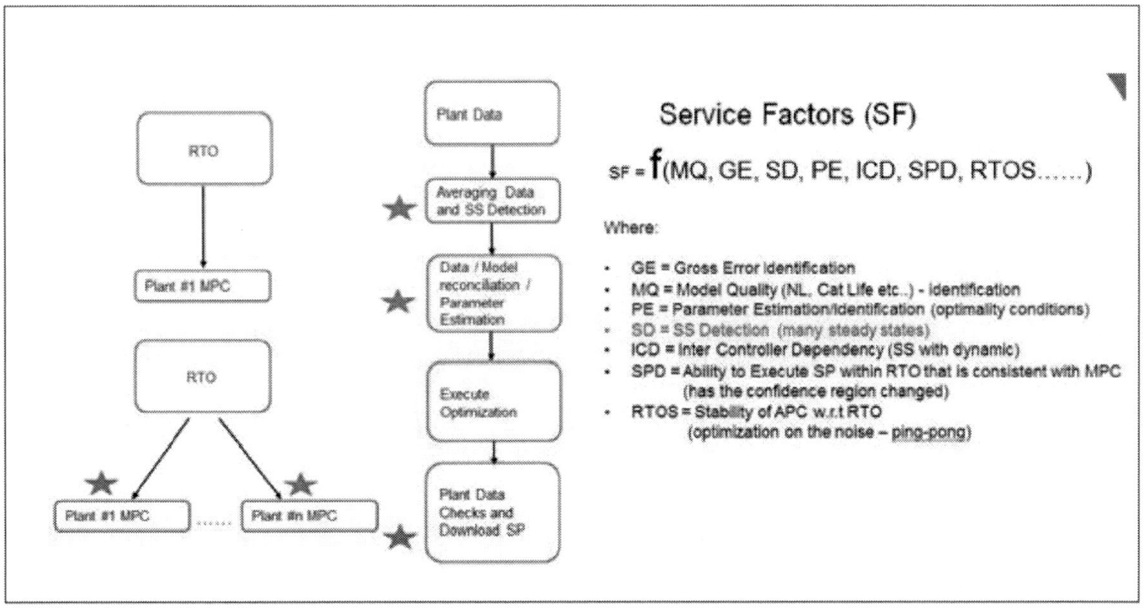

Figure-3 Simplified Product Rundown Flows and Inline/Online Blending

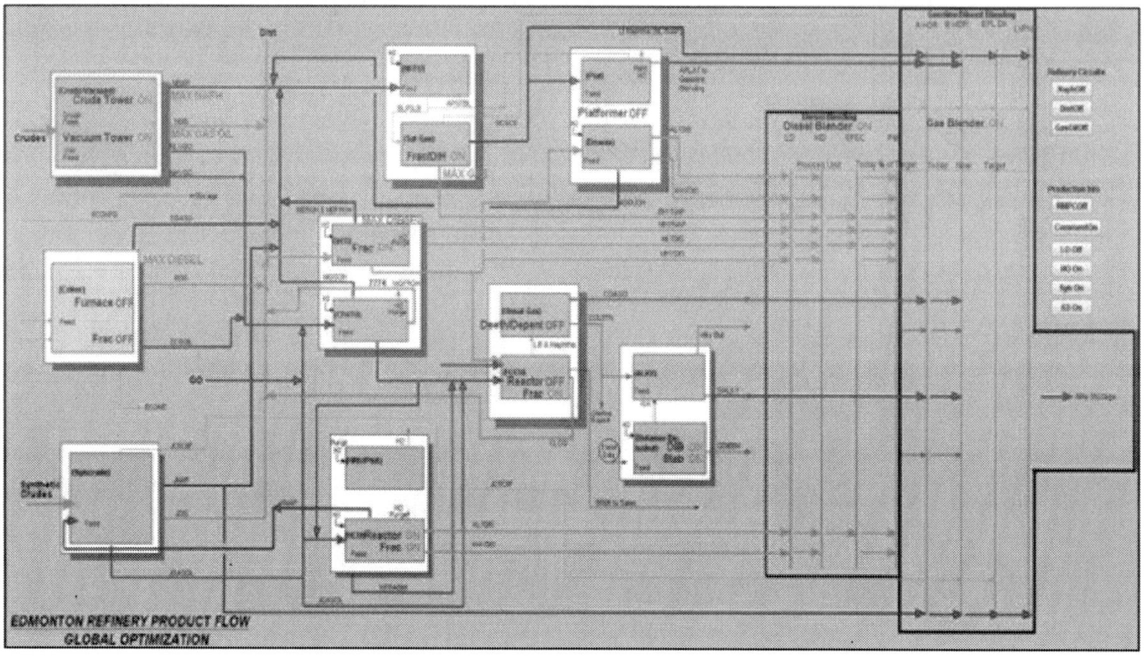

Figure-4 Diesel Inline/Online Blending directly to product tanks

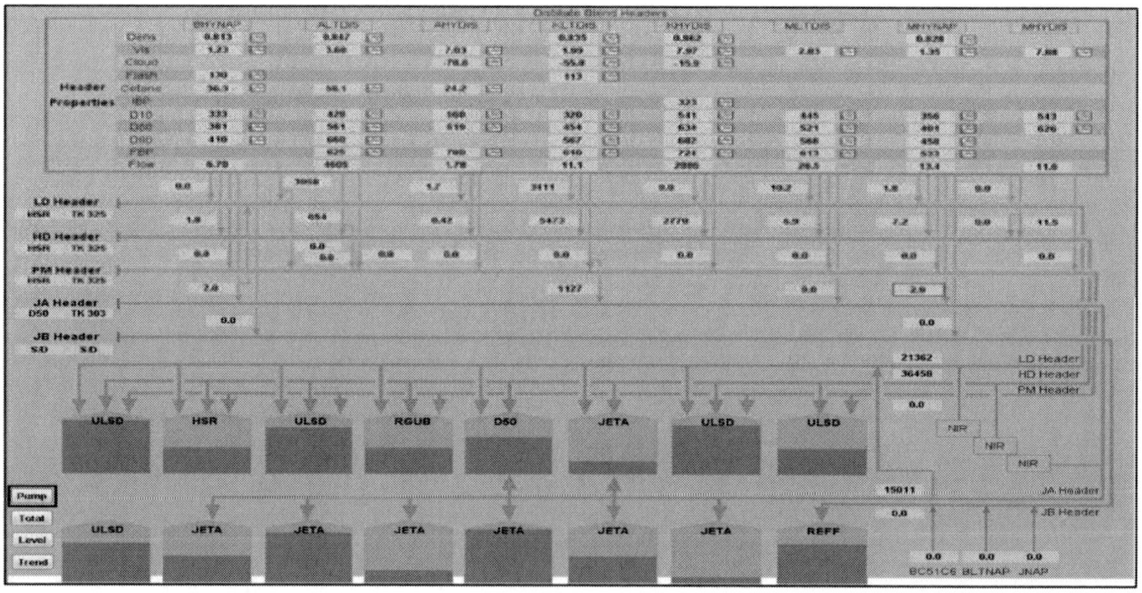

Figure-5 Diesel Inline/Online Blending Controls.

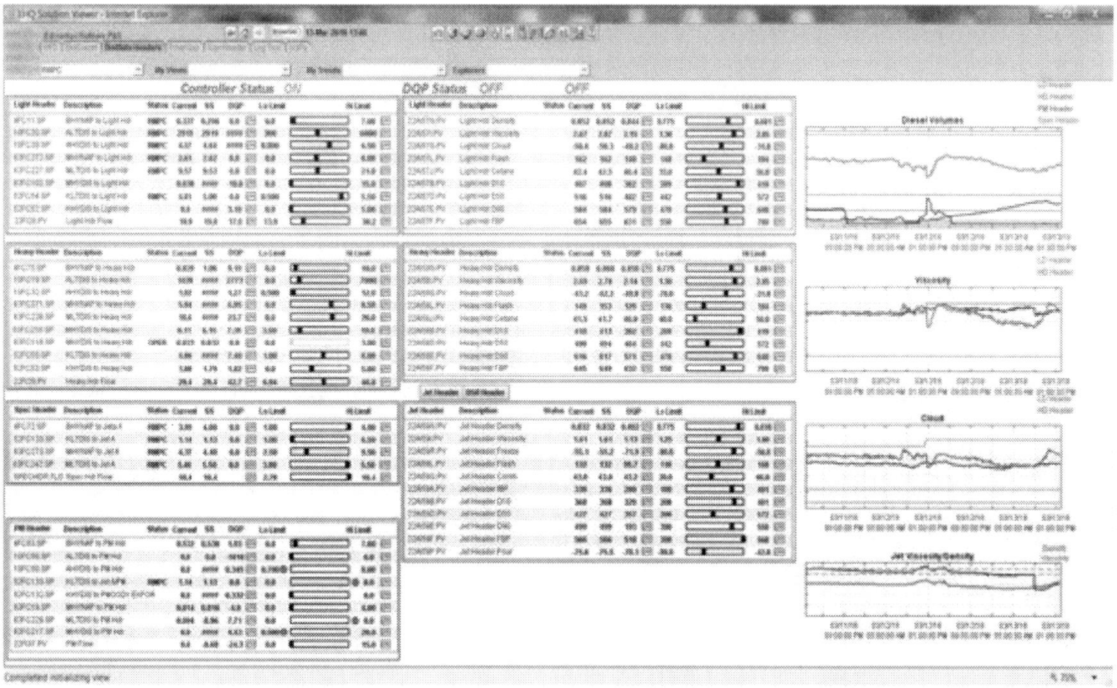

Figure-6 Issues related to Steady State detection with multiple processes in series

Process noise and steady state 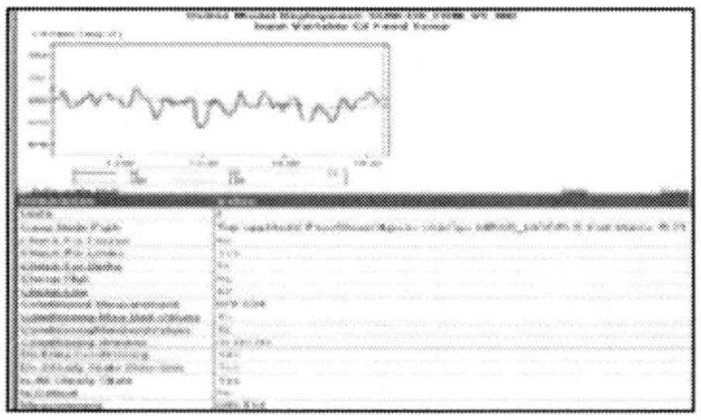	Higher process noise 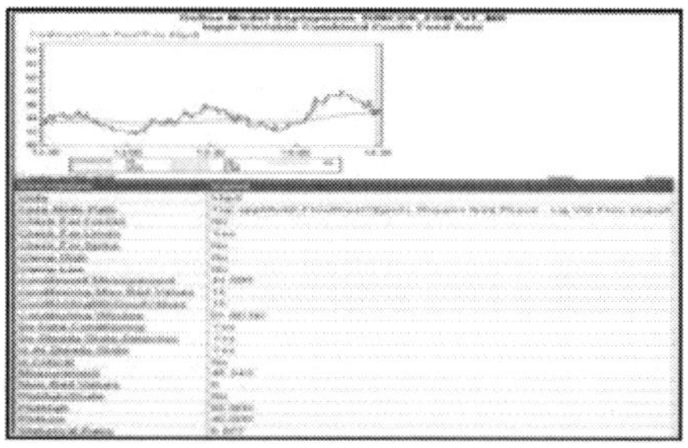
Steady state prediction lags 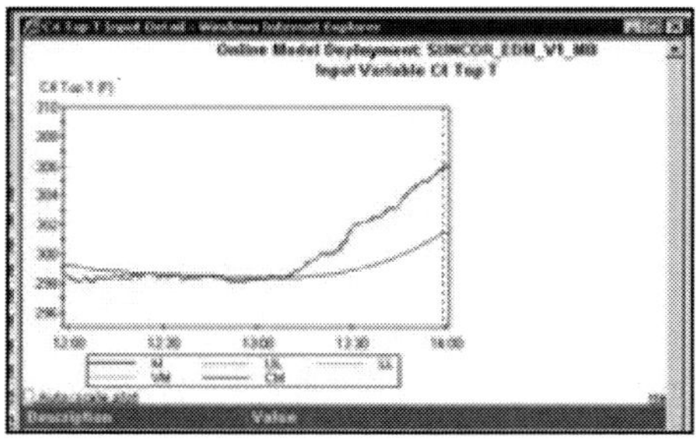	Loss of opportunity 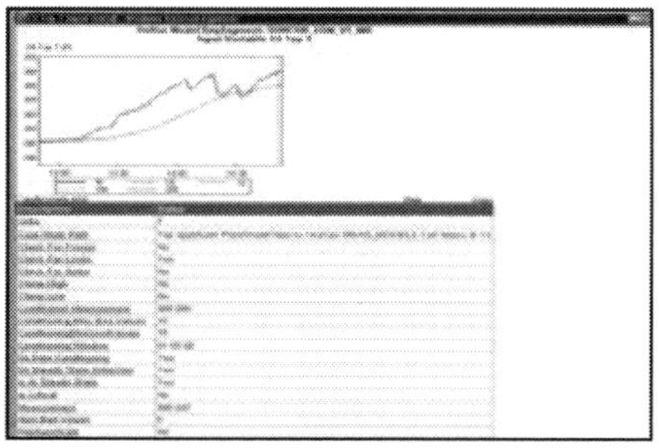

Figure-7 Fixed bed reformer catalyst cycle time yield and deactivation model

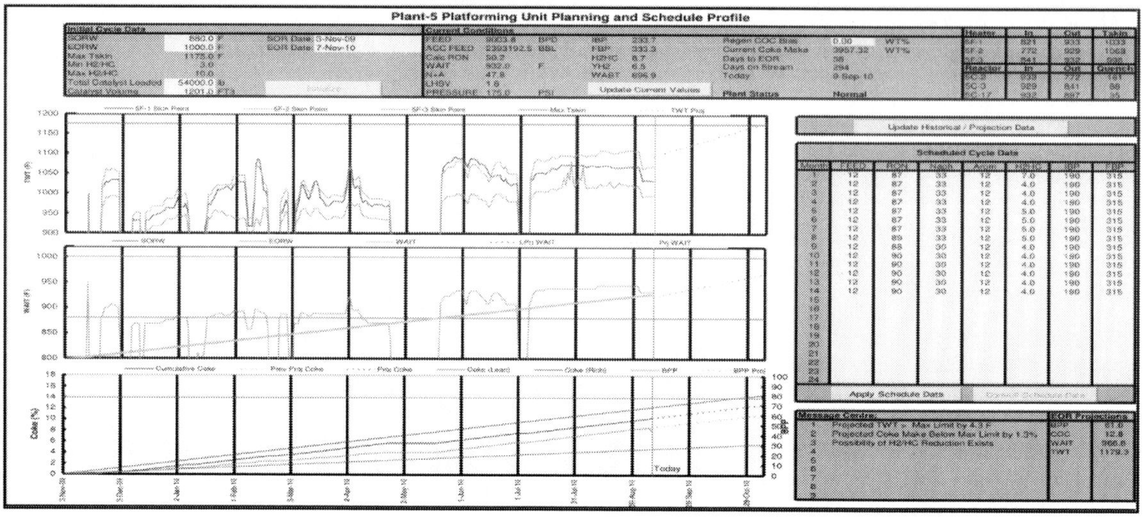

Figure-8 Refinery wide DQP connectivity Structure

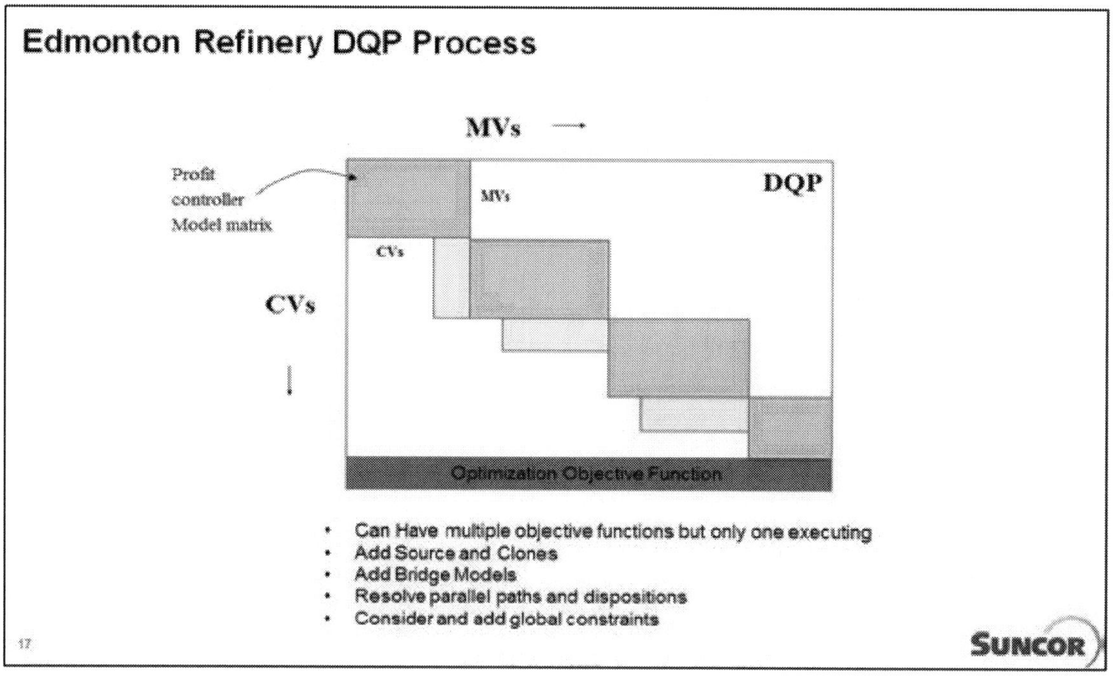

Figure-9 Refinery wide DQP connectivity Structure

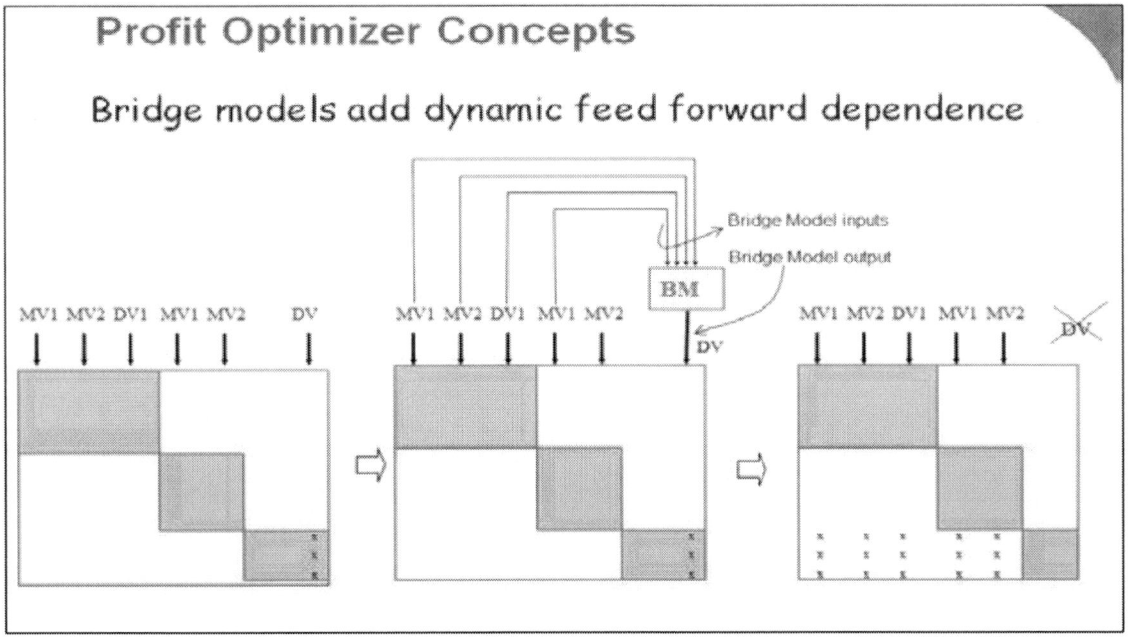

Figure-10 Refinery wide Optimization Structure and Non Linear Process Gain Mapping

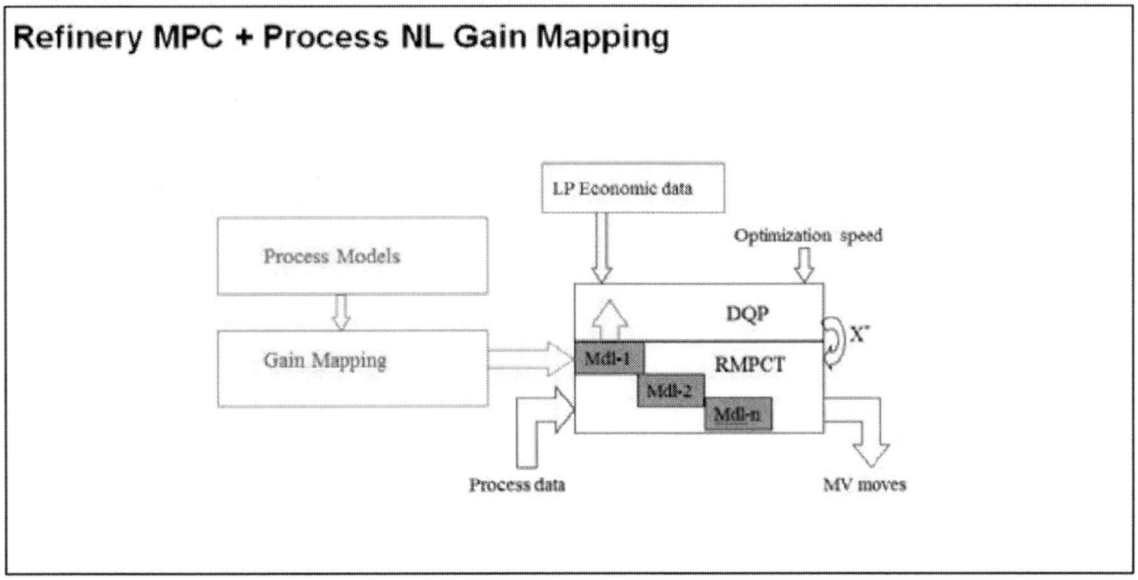

Figure-11 Refinery wide Optimization Prototype Structure

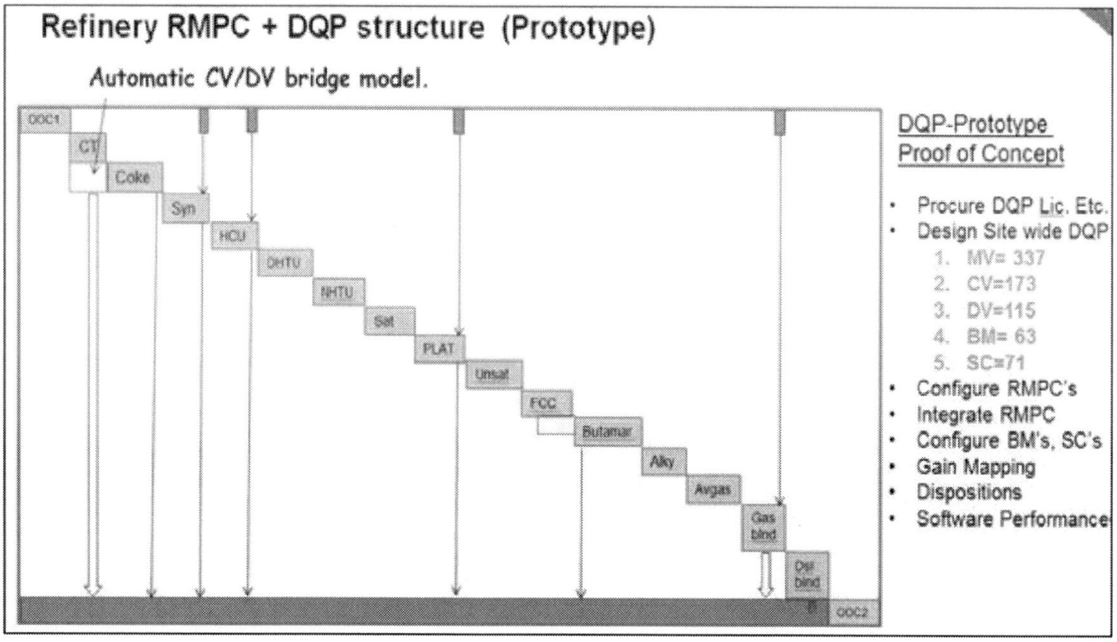

Figure-12 Refinery wide Optimization Structure

Refinery DQP Prototype

DQP ID #	Plant ID	Controller ID	CVs	MVs	DVs	Source CV	Source MV/DV	Cloned DVs	Clone ID	Feed 1	Feed2	CTRL OFFSET	Comments
1	OOC1	00IN1	1	6	0		6						Purpose of this OOC is to make indexes available to DQP
2	Crude	61IN3	29	12	6	1	12			MV1.1		05	CDU MV1-12 cloned as OOC2 DV1-12
3	Coker	69IN1	25	12	4		1	1		DV2.1	DV2.2	10	DV1 (same as CDU CV29) will be CV/DV BM. DV2 is cloned to OOC2 DV13
4	Syncrude	31IN2	12	4	3			1	51FC26_SP	DV3.1		15	Feed DV1 is cloned to OOC1 MV1 Feed DV2 (outlet from HCU) - currently ignored
5	HCU	52IN1	24	5	1			1	52FNI1_2_PV	DV4.1		20	
6		52IN2	9	4	3							25	In DCS graphics HCU has 2 feeds 51FX20A & 51FX23A, they must add up to feed DV ? Confirm.
7	DHTU	63IN1	12	12	1					DV6.1		27, 77	DV6.1 is distillate feed
	CFHTU												Gasoil feed would be an MV
8	NHTU	2IN1	10	4	2					DV7.2		35	DV7.2 is hot charge, DV7.1 is cold charge.
9	SAT GAS	4IN1	10	4	1					DV8.1	?	40	DV8.1 is feed from NHTU. There is a second feed from Platformer
10		4IN2	10	3	3							45	
11		4IN3	17	5	6							50	
12	PLAT	5IN1	17	5	2			1	5FC2.SP	DV11.2		55	source is MV 19.28
	DEWAX												No controller
13	FCCU	8IN1	13	9	3	1	2			DV12.3		60	There are 3 feeds to FCCU, the combined feed is DV12.3
14		8IN2	12	6	4			3				65	
15	Unsat Gas	9IN1	11	3	3					DV14.1		70	DV14.2 is also labeled as feed
16	Butamer	11IN1	15	6	4		1			MV15.4		75	
17	ALKY	12IN1	5	3	2					DV16.1	?	80	There is a second feed
	Avgas												No controller
18	Gasoline Blending	22IN3	22	15	5		15	3		*		85	Modified 22IN3 - mv9 is not used, mvs12, 13, 14 and DV #5 added
19	Diesel Blending	22IN6	55	28	0		28			*		13, 38, 63, 88	all MVs are feeds
20	OOC2	99IN1	29	29	61			61				95	Purpose of this OOC is to make indexes available to DQP and to add additional hdl constraints
Σ			337	173	115	2	63	71					

Figure-13 Gasoline Rundown RVP Controls

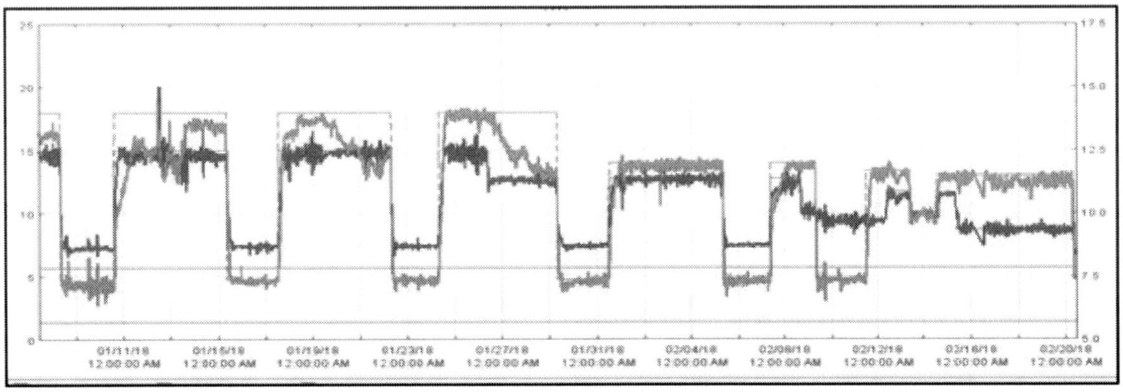

Figure-14 Gasoline Inter header optimization

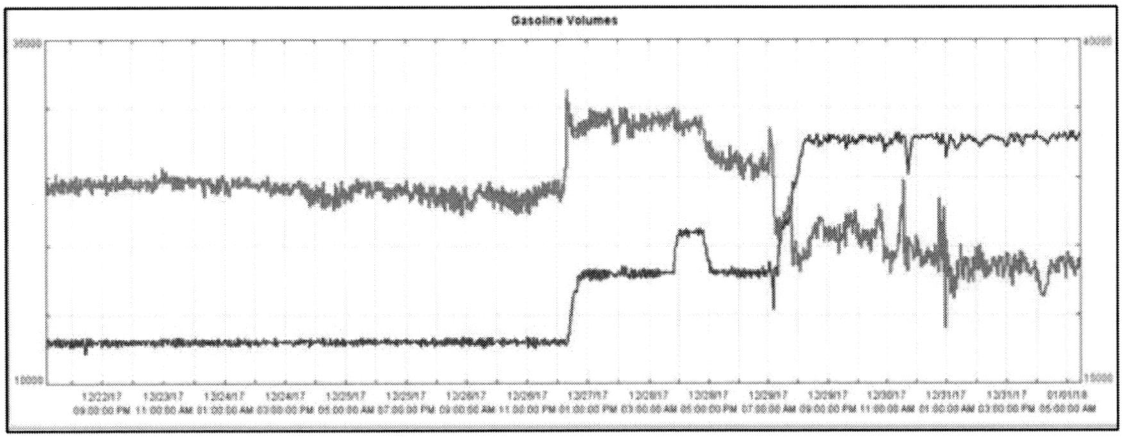

Figure-15 Gasoline Inter header optimization

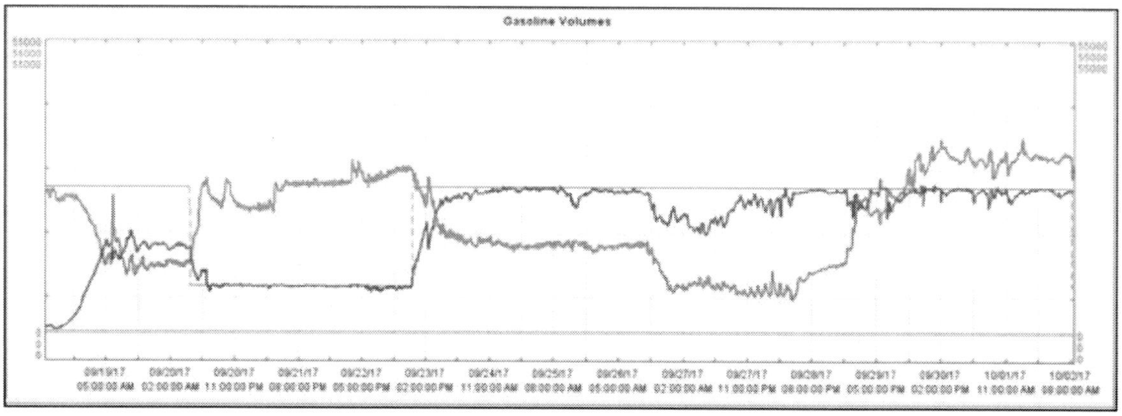

Figure-16 Naphtha/BC5C6 PVO trade off

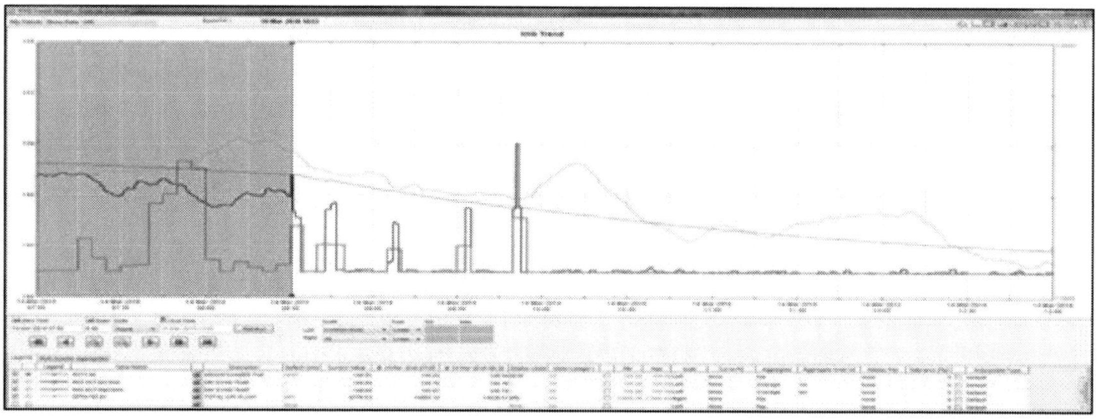

Figure-17 Naphtha/BC5C6 PVO trade

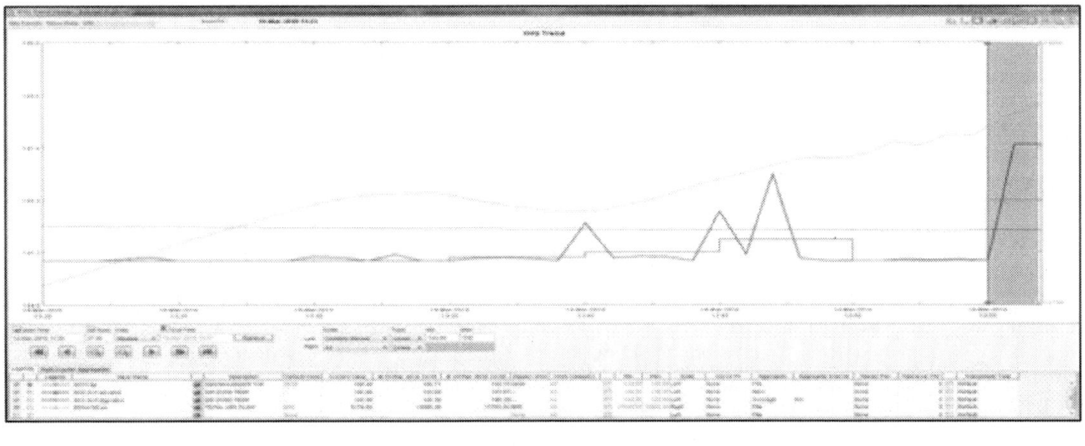

512851 Decision Automation for Lubes and Asphalt Production Scheduling Using MILP with Sequence-Dependent Switchovers

Abstract:

Hydrocarbon fractions for lubes and asphalts are produced in vacuum distillation units (VDU) by feeding reduced streams or atmospheric residue from crude-oil distillation units (CDU). The light and heavy vacuum distillates (gasoils) of the reduced crude-oils contain paraffinic, naphthenic and aromatic compounds of around 25 to 45 carbon atoms. To transform these streams into lubricant base oils, further steps of processing in solvent extraction, dewaxing and hydrotreating units are included in the process network. The VDU bottom or residue can be transformed into fuel-oils and asphalts by diluent addition to specify certain properties such as viscosity index.

In the production of lubes, the VDU downstream units remove from the lube streams aromatic compounds and long chain paraffins (waxes) using solvent extraction processes for further hydrotreating. Variations in process variables to modify distillation curves of VDU distillates and the degrees of extraction and hydrotreating define the properties of the base oil such as viscosity and sulphur and saturated carbon chain concentrations [1,2]. These lube base oils are grouped into such as group I, II, & III considering their properties, where group III are the most highly processed, most expensive and highest performing of the crude-oil derived lubes. In the production of fuel-oils and asphalts, the external input of extra diluent into the refinery field to specify the viscosity and the movement and stock of such heavy fraction streams can be a bottleneck for the scheduling optimization [3].

To schedule the blocks, campaigns or conjugated modes of operation of lube base oils, we propose a mixed-integer linear (MILP) optimization using sequence-dependent setups or switchovers [3]. Other logic and logistics constraints necessary are also included which augment the formulation for an actual lube base oil production facility. The proposed multi-stage solution solves a 30 days horizon with 1-day as time-step and the better predictions on production allow opportunities of exploring the contract and spot markets of hydrocarbon streams.

[1] Abraham, J. N.; Rao, K. S. (2009). Integration of generative and evaluative models for production scheduling of lube oil plants in a petroleum refinery. Applied Mathematical Modeling, 33, 1213â€“1227.

[2] Kyungseok, N.; Joohyun, S.; Lee, J. H. (2017). An optimization based strategy for crude selection in a refinery with lube hydro-processing. Computers & Chemical Engineering, October, just accepted.

[3] Pinto, J. M.; Joly, M.; Moro, L. L. (2000). Planning and Scheduling Models for Refinery Operations. Computers & Chemical Engineering, v. 24, p. 2259-2276.

[4] Kelly, J. D.; Zyngier, D. (2007). An Improved MILP Modeling of Sequence-Dependent Switchovers for Discrete-Time Scheduling Problems. Ind. Eng. Chem. Res., 46, 4964.

513728 Automation Strategies for Crude Feedstock Optimization

Abstract:

It used to be that refineries could count on a fairly consistent feedstock. That is no longer the case with shale oil and other discounted opportunity crudes contributing to the crude diet for most refineries. The processing of these crudes has resulted in a number of challenges for refineries including asset deterioration, bottlenecks and slowdowns. This paper examines the measurement and control technologies and strategies that are available to enable refineries to avoid problems and maximize profits while successfully processing opportunity crudes. The technologies discussed allow the user to quickly and effectively characterize the crude, adjust feedstock blending, monitor for asset deterioration including both corrosion and exchanger fouling and optimize chemical injection and desalter operations, including remote monitoring with IIoT strategies. Examples of effectively utilizing automation will be provided from refineries that process several opportunity crude oils.

Improving Crude Oil Fouling Monitoring, Prediction and Mitigation Strategies in Refinery Preheat Trains

Emilio Diaz-Bejarano[1], Francesco Coletti[1,2*], and Sandro Macchietto[1,3]

[1]*Hexxcell Ltd., Innovation Hub, Imperial College - White City Campus, 80 Wood Lane, London W12 0BZ, UK.*

[2]*College of Engineering, Design and Physical Sciences, Brunel University London, Uxbridge, UB8 3PH, UK*

[3]*Department of Chemical Engineering, Imperial College London, London SW7 2AZ, UK*

* f.coletti@hexxcell.com

Abstract

Traditional heat exchanger monitoring methodologies rely on fouling resistance calculations that neglect the effects of fouling on pressure drops, and are not able to predict future performance as a function of process conditions, heat exchangers geometry and network configurations. In this paper, an improved approach to monitoring of fouling in refinery pre-heat trains that rely on rigorous predictive models is summarised and illustrated with an industrial case study.

Introduction

Monitoring of crude oil pre-heat trains in atmospheric distillation units is key to ensure profitable and environmentally friendly refinery operations [1]. Traditional monitoring methodologies that rely on fouling resistance calculations neglect several key factors, including the effects of pressure drops, and are not able to predict future performance as a function of process conditions. As a result, maintenance and operating decisions are not data-driven and heavily relay on the operators' experience.

In this paper, an industrial case study is discussed that illustrate how advanced models implemented in Hexxcell Studio™, a comprehensive software framework integrates solutions for monitoring, design and retrofit analysis of heat exchangers and their networks, can be used to monitor operations, diagnose causes of fouling, and accurately predict future performance and lead to improved overall economics, environmental impact and safety.

Method

Four analysis steps are applied in the case study:

1) *Dynamic Data Analysis™*: the evolution of the characteristics of the deposit is inferred directly from the plant data as a function of measured inlet conditions of temperature and flowrate, and heat duty calculated from primary measurements.
2) *Parameter estimation*: the fouling deposits grow over time as a function of local operating conditions according to deposition models. Unknown fouling parameters are estimated by fitting the models to and validating against extensive sets of plant data.
3) *Predictive cleaning scheduling and flow split control*: the predictive models in b) are used to produce improved cleaning schedules and flow split schemes by applying an algorithm with energy and economic based decision criteria.
4) *Dynamic Retrofit Test™*: an alternative retrofit design is proposed for one of the exchangers with the aim of reducing fouling and improving its performance. The retrofit is rated by "re-running history" with the new design and the fouling predictive models

obtained in b), which allows assessing the impact of the proposed geometry on fouling rates, the performance of the unit, and that of the network as a whole.

Case Study

The study focuses on the hot end of the preheat train of a refinery for which plant measurements (temperatures, flowrates) were available for ca. 1200 days of operation.

1. Dynamic Data Analysis™

The dynamic analysis is detailed analysis that uses historical plant data to provide deeper insight into the thermo-hydraulic and fouling behaviour of heat exchangers than previously possible. The analysis applied to the refinery considered allowed unveiling the time evolution of the fouling behaviour in each unit, the operating conditions inside the units in fouled conditions, the time evolution of their thermal and hydraulic performance and their relative contribution to the overall performance of the network. From the analysis, it was possible to obtain the following conclusions:

a) Mechanical and chemical cleanings were performed on recurrent basis at different frequency in different units. The efficiency of cleaning actions was estimated from the data and used later in the cleaning scheduling optimisation.
b) E01AB is the largest energy contributor to the network and was cleaned most often.
c) E04 showed a much larger fouling resistance despite having an almost identical design to other surrounding units. This was taken as indication of severe shell-side fouling.
d) Fouling severely impacts the performance of the units. The heat duty decreases down to 25% and the tube-side pressure drops increase up 250% due to the accumulation of deposits.
e) The two exchangers at the hottest end of the network, located in parallel branches, present significantly different fouling behaviour. The only difference is the higher shear in the unit with presenting lower fouling.

The reader is referred to another paper submitted to this conference for a more detailed discussion of the results [4].

2. Parameter Estimation

Once the fouling behaviour was analysed, the next step was the estimation and validation of fouling parameters to enable the prediction of tube and shell-side fouling. This step is essential to provide confidence in the accuracy of the predictions that will be used to optimise operations and test alternative designs.

The parameter estimation procedure used is detailed elsewhere [2], [3], [5]. Figure 1 shows an example of the fitting where data for a year of operations were used to fit the fouling model parameters. The remaining data, about 700 days, was predicted well within the measurement errors (±1.5% of the outlet temperature, as indicated by the green band in the figure).

3. Predictive cleaning scheduling and flow split control

The details of the simultaneous optimisation of cleaning scheduling and flow split control are provided in another paper submitted to this conference [6].

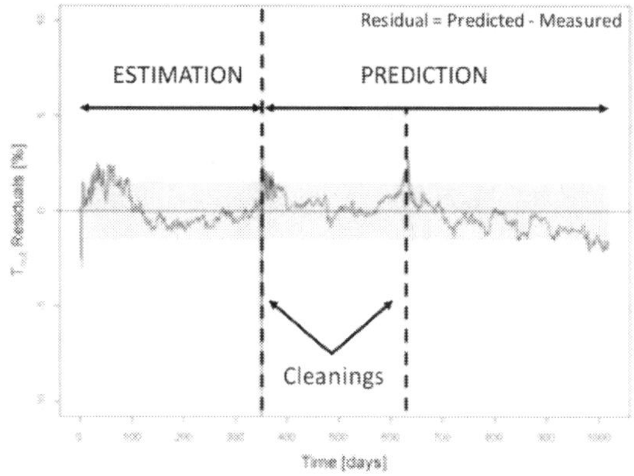

Figure 1. Model fitting and predictions for a exchangers in the network. The black line indicates the residual of the model simulation and green band indicates the uncertainty of the measurement.

The cleaning scheduling is performed by considering fixed and time-varying flow split control between the two parallel branches in the network. The results are compared to the actual cleaning schedule that was implemented in the refinery over the operating period considered. The results show that $2.7 MM savings could have been achieved compared to the cleaning schedule that was implemented in the refinery. Furthermore, the simultaneous consideration of cleaning scheduling and flow split control provides over 50% additional savings compared to the case where only the cleaning scheduling is optimised with a fixed flow split fraction.

4. Dynamic Retrofit Test™

The steps above enabled deep understanding of the network behaviour and mitigating fouling from an operational point of view. However, it would also be desirable to assess he cost effectiveness of retrofitting selected units to mitigate fouling by design. The Dynamic Retrofit Test™ is a predictive approach to test performance of a proposed retrofit design using historical plant data. The method [9,10], allows assessing how a specific heat exchanger design would have performed if in operation over a given period. It allows identifying benefits in furnace savings as a result of reduced fouling in the retrofitted exchanger taking into account complex network interactions.using this method, a retrofit design was proposed for E05AB, located at the hottest extreme of one of the branches. This exchanger was chosen because: a) it was the second greatest contributor to heat recovery in clean or low-fouling conditions; b) it underwent severe fouling, which quickly decreased its heat duty and increased the tube-side pressure drop, heavily impacting the overall performance of the network; c) the equivalent exchanger in the other branch, designed for higher shear, was presenting significantly less fouling build-up (observed at Step 1), which indicated that a high shear design on E05AB could be indeed beneficial.

A re-design moving from 4 tube passes to 2 tube passes with smaller diameter was selected for testing with Dynamic Retrofit Test™. History was re-run with the new design for the same operating period in Figure 1. Figure 2a shows the comparison of the shear stress with the proposed design (red, dashed line) and the original design (black, continuous line). The new geometry significantly increases the shear stress under clean conditions, which decreases the rate of deposition. The consequence is greater heat recovery for most of the operation period, as shown in Figure 2b. The simulations are performed for the entire, interconnected network of exchangers, allowing the assessment of the impact of the retrofit on the inlet temperature to the furnace (CIT) and, consequently, of the fuel and energy savings achieved. On average, the proposed retrofit increases the heat recovery in the exchanger by 3 MW, leading to 1.6 MW increase in the overall heat recovery of the network which translate in a CIT increase of 3.5°C. Economic savings are estimated in excess of $0.8 MM/yr.

Conclusions

This paper has shown, through a comprehensive industrial case study, how advanced models implemented in Hexxcell Studio™ can be used to monitor fouling, performance an operating conditions in fouling, fit deposition models and accurately predict future performance, and lead to improved maintenance activities, operations and retrofits with substantial economic benefits and reduced environmental impact.

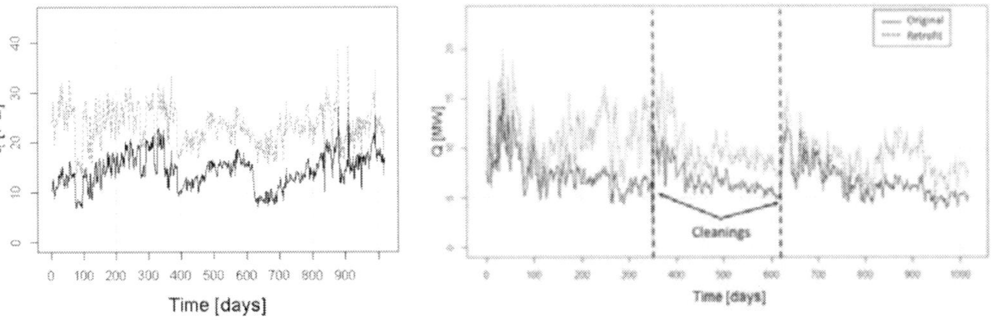

(a) (b)

Figure 2: Comparison of original design (continuous black line) and proposed retrofit design (dashed red line): a) tube-side shear stress in fouled conditions; b) heat duty.

References

[1] F. Coletti and G. F. Hewitt, *Crude Oil Fouling: Deposit Characterization, Measurements, and Modeling*. Boston: Gulf Professional Publishing, 2014.

[2] E. Diaz-Bejarano, F. Coletti, and S. Macchietto, "Thermo-Hydraulic Analysis of Refinery Heat Exchangers Undergoing Fouling," *AIChE J.*, vol. 63, no. 3, pp. 984–1001, 2017.

[3] E. Diaz-bejarano, F. Coletti, and S. Macchietto, "Modeling and Prediction of Shell-Side Fouling in Shell-and-Tube Heat Exchangers," *Heat Transf. Eng.*, vol. Accepted f, no. https://doi.org/10.1080/01457632.2018.1446814, 2018.

[4] F. Coletti, E. Diaz-Bejarano, and S. Macchietto, "137c: Dynamic Data Analysis of Large Scale Data to Monitor Fouling in Heat Exchanger Networks," in *2018 Spring Meeting and 14th Global Congress on Process Safety, Orlando, April 22-26, 2018*, p. Tuesday, April 24, 2018 043:30 PM-05:00 PM.

[5] F. Coletti and S. Macchietto, "A Dynamic, Distributed Model of Shell-and-Tube Heat Exchangers Undergoing Crude Oil Fouling," *Ind. Eng. Chem. Res.*, vol. 50, no. 8, pp. 4515–4533, Apr. 2011.

[6] F. Coletti, F. Lozano, E. Diaz-Bejarano, and S. Macchietto, "148a: Optimization of Refinery Preheat Trains: Predictive Maintenance and Operations Improvement," in *2018 Spring Meeting and 14th Global Congress on Process Safety, Orlando, April 22-26, 2018*, p. 2018 Spring Meeting and 14th Global Congress on Pr.

[7] E. Diaz-Bejarano, F. Coletti, and S. Macchietto, "A new dynamic model of crude oil fouling deposits and its application to the simulation of fouling-cleaning cycles," *AIChE J.*, vol. 62, no. 1, pp. 90–107, 2016.

[8] E. Diaz-Bejarano, F. Coletti, and S. Macchietto, "Impact of Complex Layering Structures of Organic and Inorganic Foulants on the Thermohydraulic Performance of a Single Heat Exchanger Tube: A Simulation Study," *Ind. Eng. Chem. Res.*, vol. 55, no. 40, pp. 10718–10734, 2016.

[9] F. Coletti. On the combined use of dynamic simulations and plant data to account for fouling in heat exchanger design. 2nd Thermal and Fluids Engineering Conference (TFEC), Las Vegas, NV, 2-5 April, 2017,.

[10] E. Diaz-Bejarano, M. Yugo Santos, Manolo García Dopico, L. Lanchas-Fuentes and F. Coletti. "The impact of fouling on the optimal design of a heat exchanger network: an industrial case study". Heat Exchanger Fouling and Cleaning, 11-16 June 2017, Madrid, Spain, 2017.

512404 Improving the Crude Vacuum Heater

Abstract:

Coking has long remained a key limit in crude vacuum heater performance. Many alternatives are available to reduce heater coking. These include dual-fired heaters, extended tube spacing (3-diameter) heaters, and velocity steam injection.

Another alternative that can be used in combination with any of the previous techniques is a cocurrent flow heater. In this heater, the passes are arranged to moderate the oil film temperatures and reduce coking rates. An analysis of the cocurrent flow heater is presented.

512536 Refinery Residue and Bitumen Upgrading:Gasification

Abstract:

The International Maritime Organization (IMO) has driven changes in sulfur oxides emissions allowed from sea-going vessels. The MARPOL agreement Annex VI formalizes these changes. The coming step is the January 1, 2020 target of 0.5 weight percent sulfur maximum as a world-wide limit. Multiple strategies are available to meet this target. Basic approaches include;

- fuel substitution, for example, switching to liquefied petroleum gas (LPG).
- fuel upgrading, for example, purchasing desulfurized residual oil.
- power plant switching, for example, conversion to electric power.
- post-combustion treating, for example, sulfur oxides removal in the exhaust gases.

These changes affect multiple players in the bunker fuel market. Refiners, blenders, traders, terminals, ship owners, and ship charters will all have to change their operations. Regardless of the final, new equilibrium of markets, many refiners will be forced to remove sulfur from their residue streams.

In previous iterations of removing sulfur oxides due to residue combustion, refiners have focused on five approaches.

- Finding alternate dispositions that donâ€™t required desulfurization.
- Discounting to make post-combustion sulfur removal attractive.
- Residual hydrotreating.
- Delayed coking and selling coke (with high sulfur) as a product.
-

This presentation covers the following:

- A brief introduction to the problem.
- Previous history with residue desulfurization (in the 1970s).
- Outline of the benefits of gasification for the refinery.
- Extension of gasification to other heavy streams (bitumen).
- Downstream value chain-integration with fuel and petrochemical products.
- Possible future directions.

512449 Best Practices for Bulk and Additive Transfers in Lubricant Blending Plants

Abstract:

The dynamics in the finished product lubricant make maintaining profit margins more difficult. In today's market, lubricant manufacturers have tighter product specifications, more product variations, higher quality targets, variable batch sizes, and shorter lead times to meet. Meeting new market requirements while maintaining day to day operations is difficult especially since many lubricant plants we're built 30-50 years ago. These older plants often have a lower degree of automation that leads to more bottlenecks, safety concerns, and off-spec batches compared to plants built today. One of the most significant areas for lack of automation is with additive transfers and transfers for bulk components. These transfers are often done manually and are prone to over treating, spills, ergonomic issues, cross-contamination, product losses, and incomplete transfers. Transfer of drummed additive components and bulk liquids are often overlooked as an opportunity area for impacting quality and reducing costs. In fact, an industry study showed that most lubricant manufactures focus on looking at cost reductions in maintenance practices, product testing, and labor costs instead of prioritizing quality improvements. The study found that quality improvement initiatives such as the degree of automation drives better performance and savings more than cost reduction initiatives. Automating transfer operations will actually drive better quality performance and cost reductions compared to current methods that are employed. In this session, we will compare the different methods used to transfer additivities and bulk components in lubricants plants and show the cost savings and operational improvements generated by lubricants manufactures from employing the best practice transfer techniques.

540073 Expanding the Role of LNG as a Shipping Fuel

Abstract:

The International Maritime Organization (IMO) has moved ahead with tighter sulfur caps on marine fuels. This may seem to be a problem removed from the mid-stream gas industry as the fuel of choice for the shipping industry is heavy bunker fuel. These tighter caps present several options for compliance; however none of them provide an easy fix for the shipping industry. Removing the sulfur from heavy bunker fuel is not an easy task and would require significant upgrades for many refineries. Removing the sulfur on the ship using and Exhaust Gas Cleaning System (EGCS) is also a costly upgrade for a ship.

Other solutions include switching fuel sources which is where the mid-stream gas industry could find an opportunity. Liquefied Natural Gas (LNG) is one possible option; it easily meets the sulfur requirements and also is seen as a win towards a reduction in greenhouse gas emissions. Another option is Methanol which has the advantage of simpler storage and handling. Methanol can be generated from many sources including natural gas.

This paper will look at the challenges for each of the options and the prospects for natural gas to become a significant player in fueling the shipping industry.

513260 Using Advance Control to Monitor Plant Performance and Economics

Abstract:

With thousands of applications and over a forty-year track record, multivariable predictive control has become the gold standard for advanced process control within the processing industries. Using a dynamic model of the process unit, multivariable predictive control can simultaneously optimize unit operation while maintaining strict control of key operating constraints. Yet as ubiquitous as these controllers are, they are largely under the influence of the control or advanced control engineer and provide little feedback to the process engineer or operations manager whose involvement is critical for continuous improvement of the plant.

This presentation discusses techniques for bridging this knowledge gap by providing operational insights and analytics using information obtained from the multivariable predictive controller. Included are techniques to monitor and assess lost opportunity costs, equipment constraint violations, and the plant operating envelope.

510712 Process Monitoring of Batch Bioreactor Process Using Optimized Data Based Model

Abstract:

Monitoring is an important aspect in industrial processes to ensure process safety, and maintain high product quality. Two important steps of process monitoring are fault detection and diagnosis. Data-based fault detection techniques such as multivariate statistical method have been widely discussed in the literature, and in this work, we have used partial least square (PLS), an input-output model to analyze the batch data [1]. In batch processes, the data is first arranged in a multiway fashion according to batches for easier access to train and test the PLS model. Most of industrial data have noise, correlated variables, and unknown disturbances, and thus we have introduced a filter by using multiscale based wavelet functions to de-noise, de-correlate and normalize the dataset [2-3]. Orthogonal wavelet functions decompose the original data signal into multiscale approximate, and detailed scale and additional statistical filter is applied to remove unknown disturbances, and then the filtered signals are reconstructed to form global signals.

Most of the industrial process are nonlinear in nature and application of linear PLS will lead to mismatch between model and process data, hence in this work we have used kernel extension of PLS technique as our nominal data-based model. Kernel partial least square (KPLS) is first trained with normal operating historical data to compute all the training score vectors. KPLS model is applied on the global scale to the reconstructed dataset to generate model residue or error, and fault detection decision is made by performing statistical analysis on the residue. In literature conventional statistical test, and test have been studied but they have been inadequate to diagnose the fault, and show inferior performance while detecting multiple faults compared to advanced statistical test like generalized likelihood ratio test (GLRT) [4]. GLRT solves composite hypothesis test, and provides an ability to analyze the multivariate data in time and also along individual variables, and thus improving fault detection performance for detecting multiple faults. As soon as the fault is detected, it is necessary to find the variable that presents the major deviation from its expected value; therefore, a contribution plot with GLRT is performed for diagnosis.

To further improve performance of KPLS model, selection of kernel function and its parameter is optimized, and in this work we have used multi-objective genetic algorithm to optimize selection of kernel function and its parameter value by minimizing missed detection rate and false alarm rate. KPLS being an input-output model, can also be used as nonlinear regression tool to predict the product concentration from online sensor measurements and thus providing an additional option to monitor the concentrations that are not available by sensor measurement. The developed model is applied to batch bioreactor process; experimental data of batch production of beta-carotene via fermentation is used to describe glucose consumption, metabolic product formation and depletion, and the beta-carotene production in the *Saccharomyces cerevisiae* strain mutant SM14, with 20 g/l glucose as the carbon source. Experiments are performed to validate the results obtained from the model; the process monitoring results by our developed model shows superior performance compared to conventional fault detection techniques.

Keywords: KPLS, GLRT, wavelet function, fault detection, batch processes, bioreactor.

References:

1. MacGregor, J.F., and T. Kourti. "Statistical Process Control of Multivariate Processes." Control Engineering Practice 3, no. 3 (March 1995): 403–14.
2. Bakshi, Bhavik R. "Multiscale PCA with Application to Multivariate Statistical Process Monitoring." American Institute of Chemical Engineers. AIChE Journal 44, no. 7 (July 1998): 1596.
3. Botre, Chiranjivi, Majdi Mansouri, M. Nazmul Karim, Hazem Nounou, and Mohamed Nounou. "Multiscale PLS-Based GLRT for Fault Detection of Chemical Processes." Journal of Loss Prevention in the Process Industries 46 (March 2017): 143–53
4. Mansouri, M., Nounou, M., Nounou, H., Karim, N., 2016. "Kernel pca-based glrt for nonlinear fault detection of chemical processes." Journal of Loss Prevention in the Process Industries 26 (1), 129–139.

513182 Retrofit Strategy for Heat Exchanger Networks

Abstract:

Numerous heat exchanger network (HEN) retrofit design methods have been proposed over the years, either based on topology modifications, heat transfer enhancement or a combination of both methods. However, selecting the most cost-effective retrofit strategy for a given network remains a great challenge.

In this work, the right balance between complexity of the retrofit solution, cost and benefit regarding energy recovery is sought. As such, the proposed retrofit strategy considers the application of heat transfer enhancement first, which is a low-cost retrofit option. Then structural modifications such as resequencing, stream splitting, adding a new match to create a loop or a path are explored in a step-wise approach. Finally, the third option (once enhancement and structural modifications are explored), combines the benefits of enhancement and modifications to the network to attain high energy recovery at reduced capital investment.

With heat transfer enhancement, a systematic retrofit methodology will be presented. The methodology can evaluate the feasibility of applying heat transfer enhancement techniques for a given HEN, identify the best location to apply enhancement, deal with downstream effects from the application of enhancement, and consider the effects of enhancement on pressure drop.

In general, a limitation with the application of enhancement is its effect on pressure drop together with the constraint in the degree of enhancement a heat exchanger can provide. These two factors influence the energy recovery that can be achieved. Therefore, in cases where high energy recovery is required, the next best retrofit strategy is to explore performing structural modifications. This work presents a new step-by-step retrofit approach (Pinch Retrofit Method) for performing structural modifications. This new approach is based on analysing the key features of an existing network as such providing insights into identifying the best series of structural modifications for a given network. This ensures that fewer modifications are required to achieve a degree of energy saving, and as a result is more cost-effective.

The next retrofit strategy considered is one that combines the benefits of heat transfer enhancement and modifications to the network. This can be achieved either by performing structural modifications and using heat transfer enhancement as a way of reducing the cost required (in terms of additional area required) or, replacing existing exchangers with plate exchangers (e.g. Compabloc). The use of plate heat exchangers is explored in this work as it has the benefits of improved energy efficiency, reduced fouling tendency and minimal plot space compared to shell and tube exchangers.

The methods discussed for retrofit are applied to an industrial case study. Pre-assessment on the case study identified a potential of reducing the furnace energy consumption by ~18%. The network minimum approach temperature is ~34°C, but the network is constrained with an exchanger minimum approach temperature of ~3°C. Applying the retrofit strategy proposed in this work, the best retrofit option was identified, which provided the right balance between complexity, cost and benefit. The results from analysis showed that significant energy recovery can be obtained with a payback of ~2 years.

Identifying the Causes of Acute Fouling in Refinery Preheat Train

Emilio Diaz-Bejarano[1], Francesco Coletti[1,2*], and Sandro Macchietto[1,3]

[1]*Hexxcell Ltd., Innovation Hub, Imperial College - White City Campus, 80 Wood Lane, London W12 0BZ, UK.*

[2]*College of Engineering, Design and Physical Sciences, Brunel University London, Uxbridge, UB8 3PH, UK*

[3]*Department of Chemical Engineering, Imperial College London, London SW7 2AZ, UK*

* f.coletti@hexxcell.com

Abstract

Crude oil fouling is typically treated in the literature as a phenomenon involving the deposition of organic material on the thermal surfaces of heat exchangers. Little attention is paid to the effects that inorganic species have on the thermal and hydraulic performance of the heat exchangers affected. When deposition of inorganic species occurs, it is often the cause of an upset in operations (e.g. issues with the desalter) that results in episodes of rapid (acute) fouling. In this paper, a method used to detect acute fouling episodes and identifying their causes is presented.

Introduction

Fouling deposition on heat transfer surfaces is a costly, long-standing problem particularly severe in pre-heat trains of crude distillation units [1]. Experimental analysis of fouling deposits collected upon dismantling such heat exchangers shows that both organic and inorganic species are typically present. These vary in proportion depending on the crude slates processed, refinery operating conditions, heat exchanger metallurgy and dominant fouling mechanisms. It has also been shown that the heat transfer given by fouling depends on the amount of the inorganic and organic phases, their structural arrangement and stratification or layering.

This paper presents a novel advanced mathematical model, implemented in Hexxcell Studio™, capable of describing the fouling deposit layer as a multi-component structure [2]–[4]. It is then shown how this approach is promising to detect acute fouling episodes and provide guidance on possible causes related to the composition of the deposit.

Method

Plant measurements are analysed using Dynamic Data Analysis™. This methodology uses an advanced solution mode of the shell-and-tube heat exchanger model implemented in Hexxcell Studio™ [5] to calculate the properties of the deposits over time as a function of inlet conditions of temperature and flowrate, heat duty and pressure drop. This allows estimating the tube-side deposit thickness and *apparent* thermal-conductivity over time (if shell-side fouling is negligible). The apparent thermal-conductivity is defined as the average value of the thermal conductivity of a homogeneous deposit that explains the overall thermal behaviour of the exchanger. The apparent thermal-conductivity is equal to the local conductivity of the deposit only if it remains constant over time, as this implies that the material that is depositing is of the same type and its properties do not change once it is settled. In reality, however, deposits are heterogeneous structures, with sub-layers that reflect variations in the deposition history (changes in type of foulant, temperature history, etc.). A heterogeneous deposit can be detected if the apparent thermal-conductivity varies over time. The pattern observed in the time-varying signal can be used to establish the underlying causes.

Case Study

The case study focuses on a double-shell heat exchanger located downstream of the desalter in a pre-heat train of a middle-eastern refinery. The full details of the case study are provided in [6]. Temperatures, flowrates and tube pressure drop data were available for a period of over three years in-between shutdown. The compositional analysis of the deposits, collected at the end of the run, were also available.

Results

The plant data analysis allowed calculating the time-profiles of deposit thickness and apparent conductivity, shown in Figure 1 and Figure 2, respectively. The deposit thickness profile showed a falling rate shape, with an evident acute deposition episode at around 150 days. The thickness at the end of the period was of ca. 2.75 mm.

The apparent conductivity showed variations over time, indicating the presence of a heterogeneous deposit layer. The trend shows an increase during the initial 175 days, followed by a gradual, although non-monotonic, decrease until the end of the operation period. In order to explain such behaviour, the deposition processed was postulated to follow this sequence of events: i) deposition of fouling of deposits with intermediate conductivity until the acute episode; ii) fast deposition of high conductive material during the acute episode; iii) after then, slow deposition of material with low thermal conductivity, with some occasional deposition of high conductive material.

In this context, taking into account the location of the exchanger and operating conditions (just downstream of desalter, temperature < 180°C), low and high conductive deposit can be associated to deposition dominated by organic and inorganic materials, respectively. An intermediate conductivity could be due to the two types of foulant depositing in similar proportions.

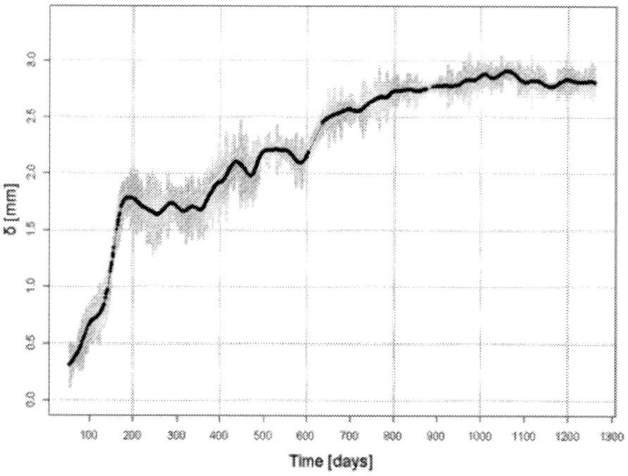

Figure 1. Deposit thickness over time (adapted from [6])

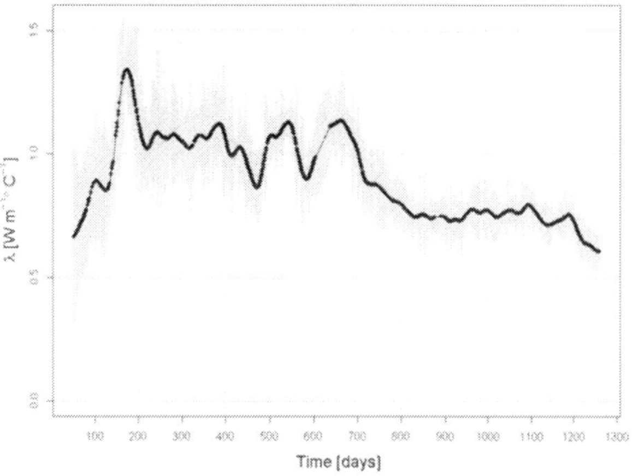

Figure 2. Apparent thermal-conductivity over time (adapted from [6])

An initial verification of this theory was performed by comparing the data analysis results to the experimental analysis of the deposits collected in the refinery. First, the apparent conductivity and thickness were used to estimate the average composition of the deposit by applying the multi-component deposit model in [3] and assuming inorganic and organic to behave as two separate phases composing the deposit. By using a Co-continuous model to describe the deposit as a heterogeneous solid, an average inorganic content of $58_{wt}\%$ was estimated. This result was in excellent agreement with the inorganic content found in the deposits, $52_{wt}\%$.

Conclusions

The Dynamic Data Analysis™ of a refinery heat exchanger allowed calculating the time-evolution of the thermal-conductivity and thickness of the tube-side deposit. This was possible thanks to the availability of pressure drop measurements, in addition to the usual temperatures and flowrates. The analysis helped detect an acute deposition episode in the thickness time profile. The trends observed in the apparent thermal-conductivity indicated that such acute episode was likely due to the fast deposition of inorganic materials, which could be linked somehow to inefficiencies in the operation of the desalter. This interpretation of the results has been supported by comparison with experimental analysis of the deposits.

References

[1] F. Coletti and G. F. Hewitt, *Crude Oil Fouling: Deposit Characterization, Measurements, and Modeling*. Boston: Gulf Professional Publishing, 2014.

[2] F. Coletti and S. Macchietto, "A Dynamic, Distributed Model of Shell-and-Tube Heat Exchangers Undergoing Crude Oil Fouling," *Ind. Eng. Chem. Res.*, vol. 50, no. 8, pp. 4515–4533, Apr. 2011.

[3] E. Diaz-Bejarano, F. Coletti, and S. Macchietto, "A new dynamic model of crude oil fouling deposits and its application to the simulation of fouling-cleaning cycles," *AIChE J.*, vol. 62, no. 1, pp. 90–107, 2016.

[4] E. Diaz-Bejarano, F. Coletti, and S. Macchietto, "Impact of Complex Layering Structures of Organic and Inorganic Foulants on the Thermohydraulic Performance of a Single Heat Exchanger Tube: A Simulation Study," *Ind. Eng. Chem. Res.*, vol. 55, no. 40, pp. 10718–10734, 2016.

[5] E. Diaz-Bejarano, F. Coletti, and S. Macchietto, "Thermo-Hydraulic Analysis of Refinery Heat Exchangers Undergoing Fouling," *AIChE J.*, vol. 63, no. 3, pp. 984–1001, 2017.

[6] E. Diaz-Bejarano, E. Behranvand, F. Coletti, M. R. Mozdianfard, and S. Macchietto, "Organic and Inorganic Fouling in Heat Exchangers – Industrial Case Study: Analysis of Fouling State,"

Appl. Energy, vol. 206, pp. 1250–1266, 2017.

[7] E. Diaz-bejarano, F. Coletti, and S. Macchietto, "Modeling and Prediction of Shell-Side Fouling in Shell-and-Tube Heat Exchangers," *Heat Transf. Eng.*, vol. Accepted f, no. https://doi.org/10.1080/01457632.2018.1446814, 2018.

509071 Improving Catalytic Reformer Heaters Â Duty Using New Technologies

Abstract:

Catalytic Reformer Units are one of the most important units and are always being run at the maximum capacity. The multi stage heater used in the unit is one of the critical equipment. It is supposed to be heating the charge going to the catalytic reforming reactor. The reactor is multi stage and the reaction is endothermic. The charge is again reheated in the first stage, second stage and third stage inter heaters. In most of the units, we find that the charge heater is fired very hard and is generally limited. The inlet temperature to the heater is generally lower than the design leading to unit being limited. Furnace Improvements has developed a split flow process for increasing capacity of these heaters by as much as 30%. We use the convection section to preheat the process instead of generating steam. This increases the heat duty of the heater at a very low cost. We are presenting two case studies illustrating substantial improvement in the heater capacity.

514236 Integration of Smart Manufacturing in the Advancement of Polymer Processing

Abstract:

In lieu of improving efficiency and embracing smart manufacturing technologies, the U.S. Department of Energy (DOE) has instituted the Clean Energy Smart Manufacturing Innovation Institute (CESMII), with Texas A&M Energy Institute leading the Gulf Coast Regional Manufacturing Center. The goals of CESMII are to (1) demonstrate at least 15% improvement in energy efficiency in real plant demonstration projects within five years, supporting a later goal of 50% improvement in energy productivity in 10 years; and (2) reduce the cost of deploying Smart Manufacturing initiatives to processing plant by 50% through the development tools and technologies. These includes developing platforms that enables machine-to-plant-to-enterprise real time sensing, instrumentation, monitoring, optimization, and control of energy. The idea of Smart Manufacturing integrates our expertise in high fidelity or advanced modeling of industrial operation, big data analytics, real time optimization and leverages high performance computing to achieve a seamless development of optimal processes and energy reduction.

The most recent advances in sampling technology and sensor development has enabled continuous online monitoring of polymerization reactions, leading to an opportunity for the application of advanced controllers. In this work, we develop model based advanced control strategies for batch polyolefin reactions via the PARametric Optimization and Control (PAROC) framework. Based on accurate linear representations of an initial high fidelity model, we formulate a Model Predictive Control (MPC) scheme that delivers a trajectory tracking problem. We solve the multi-parametric counterpart of the MPC problem (mpMPC) to determine the explicit expressions of the optimal control actions. Hence, the online computational burden of the optimization problem is reduced to a simple lookup table and affine function evaluation problem. Such an alleviation in the computational cost and ease of application allows for an opportunity for the integration of the control strategies in the Smart Manufacturing cloud platform to be used by industrial partners.

513861 Smart Manufacturing: A Must to Remain Competitive

Abstract:

Rapidly evolving technologies and products are posing enormous challenges and opportunities for chemical manufacturers. Meeting these challenges is a bare minimum for surviving in todayâ€™s dynamic global markets. Whereas to be competitive a manufacturer must be able to transform each and every rising opportunity into cash or strategic benefits.

Smart Manufacturing offers a multitude of capabilities to competitively deal with the ever increasing challenges and opportunities of chemical manufacturing.

Smart Manufacturing is a set of enabling technologies, methodologies and tools that can be deployed to improve the core business specific technologies, operations and work processes

The components of Smart Manufacturing are generic and are usually applicable to a number of different manufacturing processes.

In this presentation we will elaborate on main components of Smart Manufacturing in the context of chemical manufacturing. Proper deployment of tools such as big data analytics, process control and optimization, real-time fault detection & diagnosis, and predictive maintenance will also be described. A strategy to adopt Smart manufacturing will also be suggested.

513097 Development of an Advanced Control Philosophy for Dry Gas Combined Steam Reforming, to Operate the Plant Efficiently during Feed Composition Changes

Abstract:

Steam reforming of natural gas is one of the very important processes in the world now, as hydrogen, a key product of the reforming, continues to be in high demand for various petrochemical and chemical industries. As the gas reserves are depleting in the world, existing reforming plants are open to use natural gas from different sources. Different natural gas sources offer different compositions in the gas. This wide range of feed natural gas composition has significant effect on the performance of the steam reformers. In order to operate the reforming process at optimum conditions, precise control of reformer temperature, and steam intake are very crucial. Disturbances in the feed gas create instability in the reforming process. Furthermore, some steam reformers are operated with carbon dioxide added to the reformer feed, to have a lower hydrogen to carbon mole ratio, as required for downstream methanol or plasticizer alcohols processes. These dry gas combined steam reformers are operated at higher temperatures and lower steam to carbon ratios than that of pure steam reformers. Hence, frequent feed composition disturbances imply greater in-stability in the dry gas combined steam reformers in terms of reformer performance, reformer temperature, methane slip, fuel gas consumption and excess air in reformer box.

The existing control philosophy of the reformers cannot cope with the feed disturbances and can be effected by big swings in the reformer temperature, methane slip, fuel gas and steam & carbon dioxide flows in case of dry gas combined steam reformer. These disturbances may also have adverse effects on the plant efficiency, capacity, safety and reliability, and the plant integrity. In order to have a stable operation and cope with these disturbances, most of the existing plants are opting for advanced process control possibilities like Aspen DMC+ etc

In the paper, a new feed forward control strategy is described that can control the dry gas combined steam reforming process efficiently and effectively for a large composition variations. The basic methodology of the new control strategy is to control the steam reforming key parameters like reformer temperature, stream flow, process air, CO2 flow, and Fuel gas flow, based on the carbon count of the natural gas in a feed forward way. Additionally, a small correction to the individual manipulators is done in a feedback control philosophy. This feed-forward & feedback control philosophy enables the operating plant to react to the disturbances quickly and efficiently. Hence, the plant performance disorder is kept minimum and huge swings in the parameters are avoided during the feed disturbances. Also, with this new control strategy, the variations in the plant parameters are reduced and stable and steady production of the hydrogen can save the energy about 0.5 GJ/ton of H2 produced from the reforming process. More than the energy, the process operates more steadily and downstream processes that depend on the reforming process operate steadily as well. This developed strategy is tested and implemented and has improved the plant reliability and integrity.

514266 cMIST Compact Inline Gas Treating Technology

Abstract:

cMIST™: Compact Inline Gas Treating Technology

Shwetha Ramkumar, Norman K. Yeh, Don P. Shatto, Chuck J. Mart

ExxonMobil Upstream Research Company

Background

The natural gas industry is increasingly pursuing compact and low weight processing technology to meet the technological and economic demands of offshore, onshore, remote, and challenged gas processing. ExxonMobil Upstream Research Company has developed the Compact Mass transfer and Inline Separation Technology (cMIST™) gas treating system, a novel, compact and low weight processing technology platform, to achieve process intensification in processing facilities.

Aims and Methods

The cMIST™ system can replace conventional absorbers with in-line equipment for mass transfer and separation and can be applied to gas dehydration, selective H_2S removal and low level CO_2 removal. The first application of the technology, which is dehydration of natural gas to pipeline specification, has been qualified and licensed to Sulzer Chemtech for deployment on offshore and onshore facilities. It was demonstrated using a skid-mounted test unit in the Woodford trend area of Oklahoma, USA at an XTO production facility.

The selective H2S removal application is currently in demonstration phase at an XTO gas plant in Teague, TX.

Results and Conclusions

The cMIST™ dehydration system was tested to examine the dehydration capacity, operability, and reliability and it achieved pipeline specification over a range of process conditions and during sustained operation. The cMIST™ system for selective H_2S removal is currently being demonstrated with MDEA solvent and is achieving pipeline specification with significantly lower CO_2 pickup compared to a conventional contactor over a range of process conditions.

The cMIST™ system enables 50% system weight reduction and 50 - 80% absorber footprint reduction along with significant capital cost savings. The cMIST selective H2S removal system has added benefits of lower solvent circulation rate and the potential for elimination of the Acid Gas Enrichment (AGE) unit. The technology is highly modular to allow for simple transportation into remote, challenged, or offshore environments and provide installation configuration flexibility. The technology is also beneficial for debottlenecking and capacity enhancement applications.

513675 Novel Cryogenic CO_2 Removal- Operating Conditions and Economic Analysis

Abstract:

Natural Gas (NG) pretreatment systems have been simulated to remove more than 99% of the CO_2 from NG. Steady-state simulations with Aspen Plus™ illustrate the effects of different parameters on CO_2 capture percentage. Vapor mole fractions of CO_2 and methane also have been investigated by changing the pressure at constant temperature. Increasing pressure increases CO_2 capture and hence methane mole fraction in the vapor. Consequently, mole fraction of CO_2 in vapor decreases. As pressure increases, the methane in the vapor also condenses, creating more liquid in which solid CO_2 can dissolve. Increasing methane increases both CO_2 and C_1 in the vapor stream and decreases the CO_2 capture percentage. Increasing ethane and propane increases both CO_2 and C_1 in the vapor and decreases the CO_2 capture percentage. An experimental bench-scale apparatus verified the simulation results. Exergy loss and heat exchange efficiency have been calculated and equipment sizes and cost estimates are complete. The results confirm the low exergy losses of 39.6 kWh/ kg treated NG for simulation and 44.8 kWh/ kg treated NG for bench-scale apparatus. High heat exchanger efficiency of +90% has been achieved. High pressure and low methane in NG composition results in high CO_2 capture, though the capture efficiency reaches a maximum at intermediate pressure.

512429 Dynamic Intensification of Separation Processes

Abstract:

Process intensification (PI) has captured significant attention as a set of innovative design principles that can dramatically enhance process performance, reducing energy/capital cost and increasing process efficiency[1]. PI technologies usually integrate multiple coupled phenomena into one physical device to optimize mass, heat transfer and reduce equipment size. However, the advantages of PI, including reduced holdups, increased safety and more nimble operation come with some drawbacks and operational barriers. These include a reduction in the number of degrees of freedom (DOF) available for control, as well as a narrower window for optimization compared to conventional process equivalents[2].

In this paper, we will introduce a new dynamic operation paradigm that can be applied to existing intensified equipment to further enhance process performance. In particular, we study the use of periodic operation to manipulate and hijack the competing phenomena present in an intensified system to the benefit of economic indicators such as energy efficiency and yield. We refer to this approach as â€œdynamic process intensification.â€

Motivated by their large energy consumption across the chemical and petrochemical industries, we focus specifically on the dynamic intensification of distillation systems. Using the canonical model of a binary distillation column, we demonstrate theoretically the benefits of dynamic intensification. We then carry out extensive dynamic optimization and simulation studies to confirm the results of the theoretical findings and quantify the magnitude of the economic benefits available in a system of practical relevance.

References

[1] Creative Energy. European Roadmap for Process Intensification. 2007

[2] Schembecker, G.; Tlatlik, S. Process Synthesis for Reactive Separations. Chem. Eng. Process. 2003, 42, 179-189.

Top Dividing Wall Columns in Process Intensification

Manish Bhargava, Roomi Kalita

GTC Technology US LLC., Houston, USA
mbhargava@gtctech.com
rkalita@gtctech.com

Dividing Wall Column (DWC) technology provides refiners with a unique way of improving the efficiency and profitability of the traditional refining techniques. This technology, which has been in use since the 1940s, can be modified for various applications ranging from naphtha/reformate splitters to gas plants and can even be used to transform established technologies like isomerization and naphtha hydrotreating.

In this paper, we discuss the distinct advantages that DWCs can offer to improve and modify traditional gas plant process schemes with emphasis on top dividing wall columns. DWCs are often characterized by the presence of a vertical wall in a regular distillation column. When the distinctive wall is present at the top of the column, the column can be referred to a top dividing wall column.

Top DWCs offer a host of advantages, especially in gas plants which are generally operated under high pressures. Unlike a conventional distillation column, the top of these columns is segregated into two separation zones within the same shell with no intermixing between the two zones. As a result, they can be operated in separate modes and basically behave as two independent columns. Absorption and distillation can be carried out in a single column. This principle forms the basis of GT-LPG MAX® technology.

The concept of Top DWCs can also be implemented in a reformer unit gas plant and a FCC (fluid catalytic cracking) unit gas plant, wherein two or more columns operate in a sequence and under similar operating pressures and temperatures. The DWC will potentially combine the operation of two or more columns, thereby minimizing both capital and energy consumption of the whole sequence.

Top Dividing Wall Columns in LPG Recovery from Off gas

LPG (Liquefied Petroleum Gas) recovery units are an integral part of any refinery. In general, in LPG unit two or more columns remove the C_1-C_2 components along with other non-condensable gases, while concentrating the C_3-C_4 components to generate LPG. Figure 1 shows a standard design consisting of a deethanizer column followed by a depropanizer column. The deethanizer column operates at a high pressure, wherein the C_1-C_2 components are removed as the top product. The C_3 rich (or C_3-C_4) stream is removed in the next column. The recovery of LPG in this design is generally poor due to the high operating pressures and propane losses in the off-gas. In some processes, these losses are mitigated by using refrigeration in the overhead system. This increases the capital cost of the process.

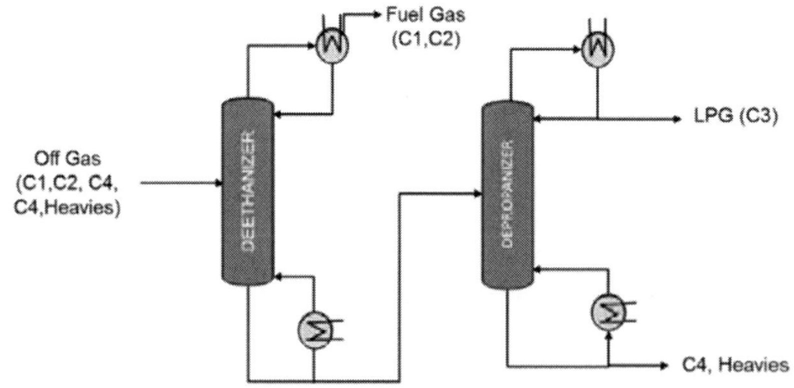

Figure 1: Conventional LPG Recovery Unit

These problems can be rectified by using a top dividing wall column. Doing so allows the top section of the column to be operated under two different unit operations, i.e., absorption and distillation. The section of the column where the feed enters operates under reboiler absorption. The other section of the column separates a light liquid product and a heavy bottom product using regular distillation. This arrangement offers the following advantages.

First, use of absorption allows the column to be operated at a lower pressure. The absorbing solvent captures the valuable C_3-C_4 components in the off-gas and moves them towards the bottom of the column. These components are then concentrated on the other side of the DWC using distillation. For feeds which contain a good amount of C_5 and heavier components, a portion of the heavy bottom product can be used as the absorption medium (Figure 2a). Using the bottoms is highly beneficial when the feed contains at least 30-40 wt% of C_5s. Table 1 gives a sample GT-LPG MAX® feed (Bhargava et al., 2015). If the C_5 concentration in the feed is low, the bottoms solvent can be supplemented by an additional heavy oil solvent from nearby processes.

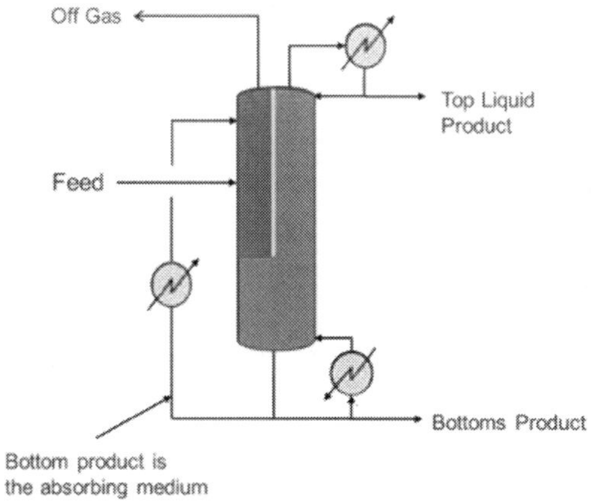

Figure 2a: GT-LPG Max® process scheme; Figure 2b: Inside view of a top dividing wall column

Table 1: Sample Feed Composition for GT-LPG MAX® Design

Product Class	Feed Composition		Product Specifications
	Components	wt.%	
Fuel Gas	Hydrogen	0.0	Minimize Propane
	Methane	0.3	
	Ethylene	0.0	
	Ethane	3.0	
	H_2S	0.5	
	CO_2	0.0	
LPG Product	Propene	0.1	Propane > 90.0 wt.%
	Propane	11.7	Ethane <1.0 wt.%
i-C_4 Product	I-Butane	8.3	Propane < 3.0 wt.% n-C_4 < 4.0 wt.%
n-C_4 Product	n-Butane	19.6	i-C_5 < 3.0 wt.% i-C_4 < 5.3 wt.%
Debutanizer Bottoms	C_5 and heavier	56.6	

Due to the low column operating pressure, water-cooled partial condensation is used in the off-gas side of the column. This is highly economical as compared to using refigeration to mitigate C_3-C_4 losses.

Additionally, this DWC design sees a reduction in energy consumption by about 10-30% as compared to traditional columns with better product recoveries (Table 2).

Table 2: Project Economics of GT-LPG MAX® Design

Variables	Units	Conventional Design		GT-LPG MAX® Design
		Deethanizer	Depropanizer	
Operating Pressure	MPaG	3.1	2.1	2.1
Reboiler Duty	MMKcal/hr	5.5		5.0
Propane Recovery	%	55.0		97.0

Top Dividing Wall Columns in Reformer Unit Gas Plants

A typical reformer unit gas plant usually has a depentanizer column, followed by a debutanizer and a deethanizer. In Figure 3, the depentanizer column removes the C_5- cut from the reformer bottoms as the top product. C_6 and heavier cuts are removed as the bottom product. In the next column, the C_5s are concentrated in the bottom of the column. The C_4 and lighter cut is then processed in a deethanizer column, which removes a LPG cut at the bottom.

The three columns in this sequence operate at high pressures. Additionally, the debutanizer and deethanizer columns see partial overhead condensation. Due to these, a major portion of the LPG components (C_3-C_4) end up in the off-gas. These components can be recovered by use of a dividing wall column configuration.

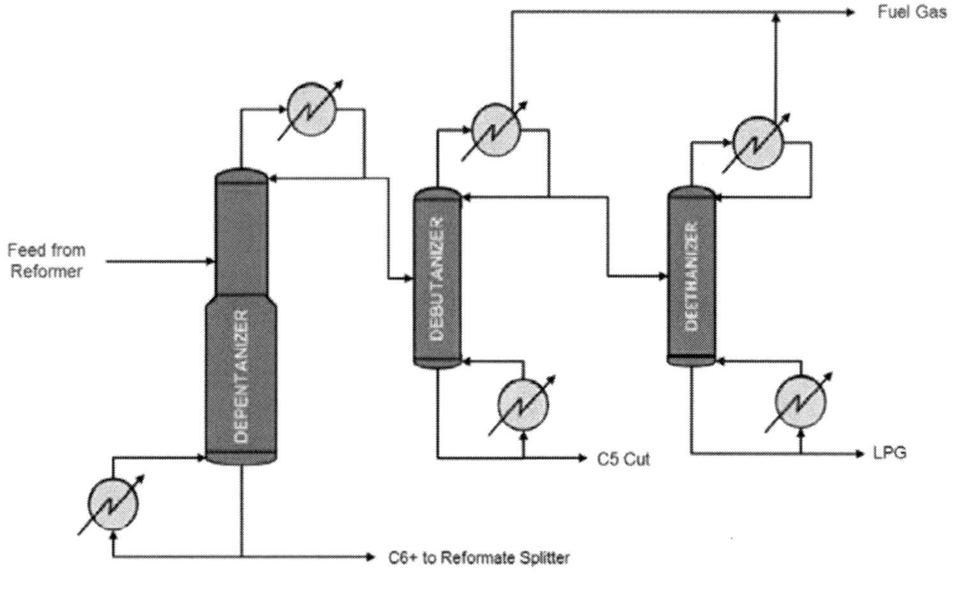

Figure 3: Conventional Reformer Unit Gas Plant Configuration

As shown in Figure 4, the operation of the deethanizer can be squeezed into the bigger depentanizer column by means of a dividing wall. The column has two distinct zones within, a pre-fractionation zone and a main fractionation zone. The pre-fractionation zone sees a combination of absorption and distillation. The lights are separated on this side as the off-gas, with the absorption medium reducing the liquid losses. On the main fractionation side, the C_5 and lighter components are removed as the other top product. Similar to a GT-LPG MAX® design, a portion of the heavy bottoms (mainly C_6-C_7) is used to strip the gas. The GT-DWC® depentanizer column is operated at a lower pressure as compared to the conventional column.

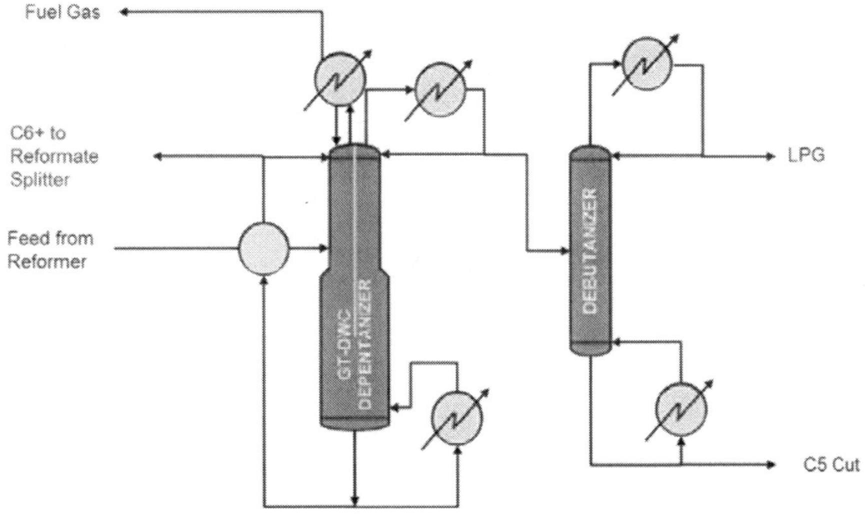

Figure 4: Reformer Gas Plant Configuration with GT-DWC® Technology

Since the majority of the lighter components is removed in the first column, the subsequent debutanizer column operates with a total overhead condenser. This column removes the LPG cut and the C_4 cut. Table 3 compares the performance of the two configurations. For the same heating duties, the DWC configuration generates better product recovery while minimizing the amount of equipment needed. The DWC configuration has a lower total installed cost over the conventional configuration.

Table 3: Performance of GT-DWC® vs Conventional Design

Parameters	Units	Conventional Design			GT-DWC® Design	
Columns		Depentanizer	Debutanizer	Deethanizer	Depentanizer	Debutanizer
No. of Trays	-	40	33	27	40	40
Operating Pressure	MPaG	0.9	1.1	3.1	0.5	1.1
Total Heating Duty	MMkcal/hr	37.1	2.2	0.5	29.0	2.7
Total Installed Cost	MMUSD	10.7			8.5	
Feed	tonnes/hr	386.8			386.8	
C6+ Cut	tonnes/hr	365.5			365.5	
C5- Cut	tonnes/hr	8.1			8.2	
Off Gas	tonnes/hr	5.8			1.7	
LPG	tonnes/hr	7.4			11.4	

Top Dividing Wall Columns in FCC Unit Gas Plants

Figure 5 shows a traditional FCC unit gas plant. The process scheme primarily consists of absorber/stripper columns, where the gases and the liquid in the feed are separated. Unstabilized naphtha (supplemented with a portion of stabilized naphtha bottoms) prevents the C_3-C_4 losses from the absorber column. The remaining C_3-C_4 are extracted by using lean oil in the sponge absorber column. Sour fuel gas and rich oil get separated at the top and bottom. The liquid product from the stripper column is processed in a debutanizer column. Sour LPG is extracted as the top product, while stabilized naphtha is recovered at the bottom.

The conventional design consists of three stages for the separation, the absorbing and stripping section followed by distillation in the main debutanizer column. Using DWC technology, the three stages can be integrated into a single column (Figure 6).

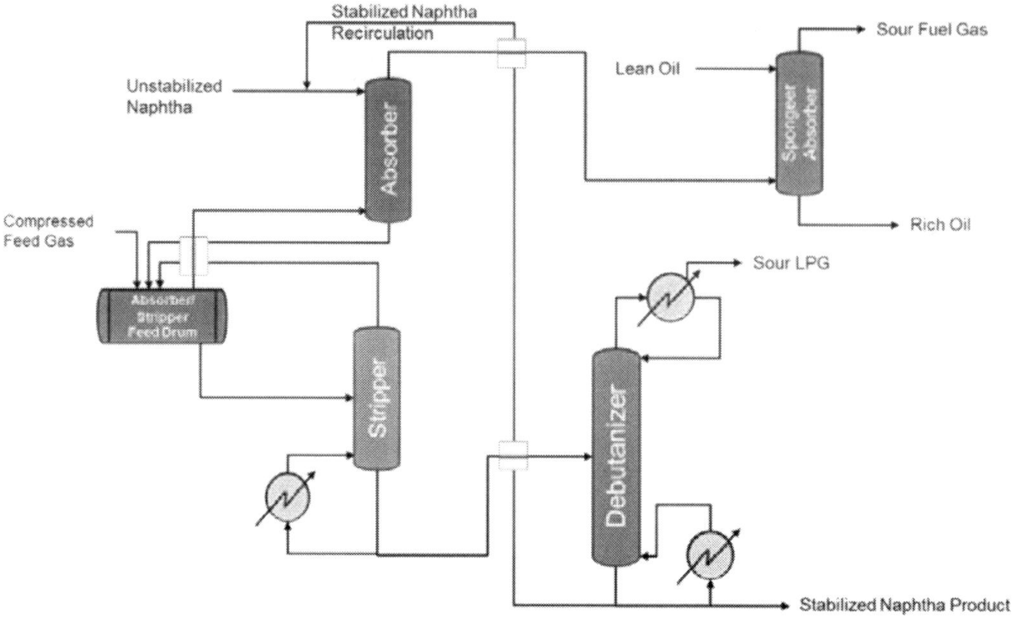

Figure 5: Typical FCC Unit Gas Plant Configuration

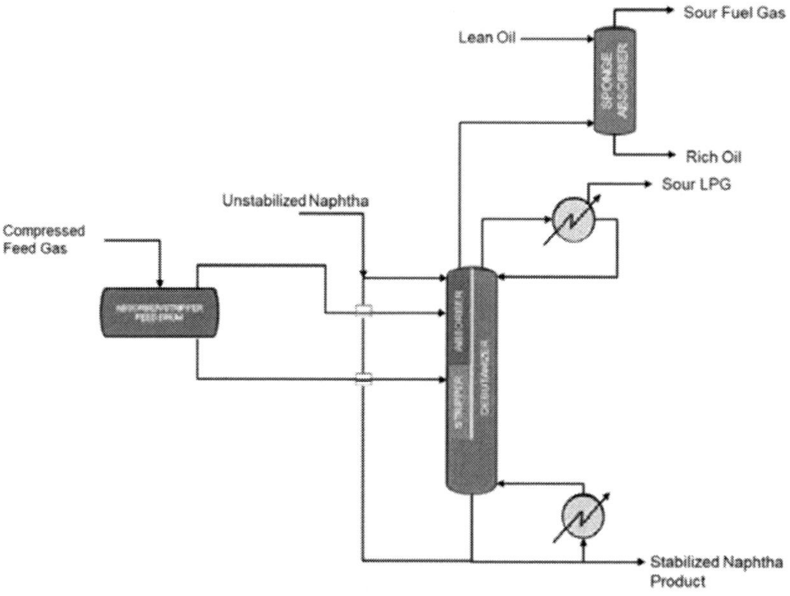

Figure 6: FCC Unit Gas Plant Configuration with GT-DWC® Design

Due to the segregation of the top halves, the two sides of the column behave as individual columns with independent overhead systems and controls. The parallel zones provide similar stages for separation in a single column. A portion of the bottoms product is circulated as the absorption medium in the section where the feed enters. This operation is similar to that of the absorber in the conventional design. The liquid portion of the feed enters below the absorption section. The C_3-C_4 components move down the wall aided by absorption. Once across the wall, these components are concentrated as the top product by rectification and stripping. The stabilized naphtha is removed as the bottom product. Table 4 gives a comparison between the two designs.

Table 4: Comparison between Conventional Design and GT-DWC® Design

Parameters	Units	Conventional Design	GT-DWC® Design
No. of Columns	-	4	2
Total Heating Duty	MMKcal/hr	9.9	7.9
Utility Savings	%	-	20.0
Feed rate	tonnes/hr	55.3	55.3
LPG Recovery	Vol %	85.1	85.1
C4 Recovery	Vol %	98.8	99.6

Conclusion

Gas Plants are a vital component of any refinery. These columns have always been operated using well-established distillation techniques. However, using the Dividing Wall Column concept, primarily top DWCs can generate a leaner and efficient system. DWCs not only lower the equipment count (columns, pumps, and reboilers' et al.), these columns also lower the energy consumption of any configuration. On occasions, these novel distillation columns can even provide better and sometimes new products as compared to the conventional columns. Integrating DWCs in gas plants can virtually change the way the refining industry improves the profitability of their gas plants in the coming years.

References

Bhargava M., Nelson C., Gentry J.C., Siddamshetti, V., 2015, Improved Distillation Efficiency, Hydrocarbon Processing (January)

512469 Improving Profitability By Solving Crude and Fuel Blending Challenges

Abstract:

The global refining industry has faced constant change in recent years most significantly around the increased availability of opportunity crudes and stricter clean fuel regulations. These changes have made blending at both ends of the refinery more challenging and even more critical to profitability. Certain blending methods and technologies are now no longer optimal for blending these new opportunity crudes or meeting these tighter fuel product specifications. In this session, we will discuss the limitations of common methodologies and technologies utilized in blending operations in today's market and share the new best practices for minimizing these new blending challenges and optimizing your profitability at both ends of the refinery.

A Yokogawa Company

TECHNICAL PAPER:

WHAT IS THE OPTIMUM GAS OIL CUTPOINT?

Written by: Al Shelton, Principal Consultant, KBC

In many refineries, there is no discussion about the optimum crude and vacuum unit gas oil cutpoints because it is standard procedure to cut as deep as possible. However, the optimum cutpoint is dependent on the refinery configuration, gas oil and resid conversion units, crude and vacuum unit designs, the price spread between light and heavy crude oils, product economics and strategic targets. With the advent of Light Tight Oil (LTO), this decision has become more complicated, because bottom-of-the-barrel economics are changing and crude/vacuum units are being operated on different crude slates than the original design basis.

Deepcut definition

The term "deepcut" generally refers to vacuum units designed to produce a 1050-1100°F TBP cutpoint between the lowest sidestream and the vacuum resid. KBC determines the cutpoint based on the vacuum column feed. Most units operate at cutpoints between 950°F and 1050°F. Depending on the resid conversion units, gas oil conversion units and relative prices of value added products and fuel oil, there may be an incentive to deepcut or cut as deep as possible. The following sections discuss some of the key points in evaluating deepcut economics and determining the optimum vacuum resid initial cutpoint and heavy vacuum gas oil (HVGO) lift.

Deepcut operation

Deepcut or the maximum cutpoint is a function of the existing equipment, feed quality and the target vacuum gas oil qualities. KBC has proven technology which has been developed during 30 years of grass root process design projects, the optimization and troubleshooting of existing vacuum units and revamp projects. Cutpoints as high as 1150°F on North Sea crudes, 1130°F on Middle Eastern crudes and 1050°F on heavy South American crudes have been achieved in KBC designed units. Successful operation has been achieved with vacuum heater outlet temperatures as high as 815°F using KBC Deepcut Vacuum Distillation Technology. In some cases, 1150°F cutpoints are possible, while cutpoints above 1050°F are routinely achievable. However, the maximum cutpoint is not always the optimum economic operation.

A Yokogawa Company

Deepcut economics

The effects of shifting vacuum gas oil cutpoints are non-linear. An incremental barrel of FCCU feed produced by increasing HVGO cutpoint will contain a higher concentration of contaminants than an incremental barrel of full range HVGO. The vacuum resid viscosity will not change linearly with incremental changes in the initial cutpoint. Because of this non-linearity in physical properties versus cutpoint, the economic evaluation of any deepcut analysis requires representative projections of the credits and debits in the affected units. Credits might include improvements in valuable gasoline and distillate yields and the elimination of fuel oil viscosity giveaway while debits might include increased operating costs in the downstream treating units, conversion units and the catalyst cycle length of FCCU feed hydrotreaters.

Deepcut benefits depend on the refinery configuration, downstream conversion unit constraints, feed and product prices. The most reliable way to calculate the benefits is to evaluate the deep cutpoint operation through the entire refinery, using a KBC Petro-SIM refinery-wide simulation flowsheet with calibrated reactor models for all downstream units and product blenders.

An LP model can be used but the economics depend on the accuracy of the downstream conversion unit yield vectors, which may be an extrapolation of the present operation. KBC experience is that rigorous Petro-SIM non-linear reactor models are more accurate than linear models for these evaluations.

KBC has completed more than 100 crude and vacuum unit projects, including over 35 vacuum unit revamps and new, grassroots process designs. Most of the design projects have been based on planning studies and profit improvement programs conducted with a refinery-wide Petro-SIM simulation flowsheet. KBC also determines the optimum cutpoints for an existing operation in a refinery simulation. Several cases are run with a range of vacuum resid initial cutpoints. The flowsheet simulation is constrained by downstream process limits such as conversion unit capacity, maximum feed metals, concarbon, FCC air blower and wet gas compressor constraints, hydroprocessing catalyst run lengths, coke production and heater firing limits. As the cutpoint is increased, an economic optimum is reached where the penalty for metals and concarbon contamination of hydroprocessing catalysts or the FCC feed offset the economic advantage for the reduction of fuel oil production or credits for running heavier crudes. This point is then used as a basis for checking a deepcut operation.

Feed characterization

Many deepcut projects are based on limited data that may not include a wide range of future crude oil feedstocks. Grass roots projects are frequently based on one or two crude slates and revamp projects may be based on one test run to define feed and product physical properties. Petro-SIM will represent new and existing operations and accurately predict intermediate and final product yields and qualities. The ability to predict product quality is due to the KBC crude assay handling system. Every refinery oil stream is represented by a minimum of 106 assay components including 24 pure components and 82 pseudo-components. Each pseudo-component can store up to 77 physical properties. The 1000°F plus heavy oils and resids are characterized by 32 pseudo-components ensuring accurate distribution of physical properties, contaminants and metals. For vacuum deepcut projects, critical physical properties are tracked by boiling range including concarbon, C7 insolubles, asphaltenes, sulfur, nitrogen and metal contaminants such as nickel, vanadium, iron and copper.

Atmospheric gas oil cutpoint and the operation of the crude column wash zone, AGO product stripper steam, bottoms stripping steam and heater outlet temperatures can affect diesel/AGO fractionation and cold flow properties which are difficult to predict. The Petro-SIM crude assay system accurately predicts cold flow properties such as pour point, cloud and freeze points. The effects of atmospheric and vacuum gas oil cutpoints are carried downstream to all hydrotreating and conversion units and the final product blenders in the refinery flowsheet.

DOWNSTREAM CONVERSION UNITS

It is important to note that the yield credits associated with vacuum cutpoint are not realized at the vacuum unit, but rather in the downstream units. Cutting deeper provides the opportunity to optimize downstream conversion unit operations and blending options. However, the benefits vary depending on the refinery configuration and product slate. KBC has conducted Petro-SIM refinery flowsheet studies for the following configurations.

- Topping or hydroskimming refinery that produces No.6 bunker fuel oil
- Asphalt refineries
- FCCU with no upstream feed treatment
- Gas oil hydrotreater / FCC unit
- Gas oil hydrocracker
- Thermal cracker
- Delayed coker
- Vacuum resid deasphalting unit
- Supercritical extraction deasphalting unit
- Complex refineries with multiple conversion units

Fuel oil refinery

Increasing gas oil recovery from vacuum resid significantly increases profitability when the refinery produces fuel oil. Recovering gas oil at the vacuum unit upgrades low value fuel oil to

FCC feed which converts the cutpoint gas oil to gasoline and diesel. Converting gas oil in the FCC results in a volumetric expansion of value added products further increasing the profitability above the gasoline and diesel versus fuel oil price differentials. Although this deepcut opportunity would appear to be obvious, refinery flowsheet cases were required to convince some clients that less cutter is required even though the vacuum resid viscosity increases because the resid production decreases requiring less net cutter. Furthermore, cutpoint HVGO is a much lower gravity, more viscous cutterstock than a lighter, less viscous distillate cutter. Whether cutting to meet gravity or viscosity specifications, blending the lighter cutter is more profitable than cutpoint Heavy HVGO.

Asphalt production
Petro-SIM refinery flowsheet cases have demonstrated the same principle for Asphalt production. It is more efficient to cut deep and blend full range HVGO to make PG Asphalt than adjust the HVGO cutpoint on the unit to meet the viscosity specification. The full range HVGO viscosity is lower than the cutpoint Heavy HVGO viscosity so less blended cutter is required to meet the viscosity specification.

FCC versus coker yields
There have been references in the literature that 10-15 wt% of the cutpoint HVGO in the coker feed will be lost to coke but this is dependent on the crude type. KBC studies indicate that the mass yield loss to coke is lower for light crudes and may be only 2-3 wt% but could be as high as 8-10 wt% for heavy crudes where the concarbon might be 6-8 wt% at a 1050°F boiling point. In either case, if the cutpoint HVGO feeds the FCC, the coke will be burned in the regenerator. It is the difference in C3+ liquid volume yields which are 102-108% in the FCC Unit versus only 60% in the Coker for full range coker feed and 80% for 1050°F cutpoint HVGO produced from heavy crude oils that determines the refinery deepcut economics.

Coker unit constraints
If the coker is coke drum limited at minimum cycles then increasing HVGO cutpoint and reducing the vacuum resid feed to the coker will not result in increased coker throughput. However, if there are hydraulic limits such as coke drum velocity, feed system pressure drop, main fractionator flooding with lighter feeds or coker heater firing limits, then increased HVGO cutpoint will facilitate overcoming these constraints.

FCC unit constraints
If the FCC feed is not hydrotreated then increasing HVGO cutpoint reduces the quality of the feed to the FCCU. If the FCC Unit is not air blower limited, increasing the HVGO cutpoint will be profitable whether producing fuel oil or in a bottoms upgrading refinery with a delayed coker or resid thermal cracker. When the FCC is not air limited and can process additional

cutpoint heavy feed then the higher metals loading is an operating cost for increased fresh catalyst makeup or equilibrium catalyst. Improved C3+ liquid volume yields and the increased gasoline and distillate yields will more than offset higher operating costs if the FCC Unit can process the incremental feed. The FCC pressure balance is complex and can be adjusted if the wet gas compressor is not limiting. Reducing regenerator pressure will unload the air blower and load the wet gas compressor. The operation is complex and KBC frequently employs optimizers in the FCC reactor model. When the FCC feed is hydrotreated the analysis is more complex.

FCC feed treatment economics

For FCC feed hydrotreaters and mild hydrocrackers increased concarbon and metals in the feed shorten catalyst run length and those economics must be considered in the refinery simulation. For example, in one KBC deepcut study, a 35°F vacuum gas oil cutpoint increase would have reduced the run length of the FCCU feed hydrotreater with the existing desulfurization catalyst. However, by changing to a higher activity catalyst, the target run length target could be maintained at the higher HVGO cutpoint. The evaluation of the increased cutpoint project had to include the accelerated deactivation for the existing catalyst and the catalyst change out costs for the higher activity catalyst.

MVGO Hydrocracker / HVGO FCC / Delayed Coker

KBC developed the optimum process configuration for a grass roots refinery with Petro-SIM and designed the new Crude and Vacuum Unit for a complex heavy crude slate. The design basis was high TAN, heavy crudes and the heavy West African crudes diluted the cutpoint metals and concarbon in the heavy South American crudes. Based on process guarantees specified by the client, the vacuum unit was designed for a 1050°F TBP cutpoint. This is an actual TBP cutpoint because the overflash slop wax is withdrawn and pumped down to the top tray of the bottoms stripper to improve gas oil lift and recovery. The stripped, net slop wax flows from the stripper into the vacuum bottoms boot so the HVGO is the first draw above the vacuum bottoms product. In this KBC design, the 5% VTB - 95% HVGO overlap, HVGO FCP and VTB ICP are meaningful indicators of deepcut performance.

The vacuum column could operate at higher than design HVGO final cutpoints and VTB initial cutpoints. Although KBC ran hundreds of cases over a 7 year period for the conceptual design, licensor selection, process design, detailed engineering design, evaluating the startup crude slate and optimizing new operating crude cases, the optimum HVGO cutpoint was consistently 1025°F or less. The KBC designed, 3 draw, deepcut vacuum column with bottoms stripping steam was capable of 1050°F to 1100°F deepcut operation but the downstream conversion unit yields in the refinery flowsheet showed that lower cutpoint operation was more profitable.

The vacuum column is designed to produce LVGO meeting USGC and South American Diesel endpoint specifications, MVGO with a 1000°F maximum endpoint, 0.20 to 0.30 wt% concarbon and low metals for Hydrocracker (HCU) feed plus a Heavy HVGO that is blended with excess MVGO for FCC feed. VTB plus stripped overflash feeds the Delayed Coker and the HCGO is processed in the Hydrocracker with FCC LCO and VDU MVGO. Although the crude slates and operating conditions changed, deepcutting the HVGO was not profitable primarily due to the conversion units' yields. Over a range of operations, the Hydrocracker C3+ liquid yields were 110-120% compared to 102-108% in the FCC and less than 100% in the Coker. Although the cutpoint HVGO yields are lower in the coker than the FCC, the olefinic HCGO readily cracks in the Hydrocracker. Overall refinery liquid yields were higher with heavy cutpoint H-HVGO feeding the coker to produce low concarbon HCGO feed to the Hydrocracker rather than the FCC Unit which could became air blow limited without an upstream feed hydrotreater.

Deasphalting Unit (DAU)

For conventional deasphalting units and both 2 stage and 3 stage supercritical extraction units, high HVGO cutpoints at the vacuum unit do not increase gas oil recovery if the DAO feeds the GOHT or FCC Unit and there are no DAU capacity limits. HVGO does not contain asphaltenes and resins so the cutpoint HVGO is recovered in the DAO. For a 3 stage unit, yields can be predicted for full range or cutpoint feeds by laboratory analysis. Pentane insolubles will predict the DAO yield, the incremental hexane insolubles will indicate Resin yields and Heptane insolubles measure the Asphaltene yield. Unless there are resid entrainment problems in an existing vacuum column, there will be no heptane insolubles or asphaltenes in HVGO.

Deasphalting Unit (DAU) / Delayed Coker

KBC recently conducted refinery case studies for a facility with an intermediate supercritical extraction unit where the resin and asphaltene feed the coker and DAO is treated in the FCC Feed Hydrocracker. Determining the optimum vacuum resid initial cutpoint is complicated because one vacuum unit is a deepcut design with bottoms stripping but the old vacuum column is a dry operation which limits cutpoint. The newer deepcut VTB feeds the coker so maximizing cutpoint at that unit is profitable because the coker is not drum limited but the unit is constrained by main fractionator flooding. For lighter crude slates, higher cutpoints at the old unit where the VTB feeds the intermediate deasphalter did not improve yields. However, when processing heavy crude slates, VTB production exceeds the deasphalter capacity and some VTB must be fed directly to the coker. In this case, deepcutting at the old vacuum unit is profitable but the dry operation limits cutpoint so in some cases it is beneficial to recycle slop wax to the vacuum heater to improve gas oil lift from the net slop wax blended into the VTB. KBC has evaluated slop wax recycle for new unit designs and this configuration is not justified for new units with bottoms stripping steam. However, in this facility, several unconventional operations were beneficial for different crude slates and operating scenarios. It

A Yokogawa Company

would be difficult to analyze the operation of more than 8 process units without a refinery simulation.

VACUUM UNIT DESIGN

Most refineries have more than one crude and vacuum unit and as in the case in the previous section the design and operation of the vacuum column may be different. The ability to deepcut depends on the vacuum unit design. Column diameter is the ultimate constraint for any vacuum unit since wash zone vapor velocities limit throughput and gas oil lift. The wash zone C-factor limits depend on the type of packing but these constraints frequently limit increasing flash zone and vacuum heater outlet temperatures or reducing flash zone pressure to increase cutpoint.

Vacuum column wash zone

There are several objectives in the design of a wash section, some of which are conflicting. Fractionation should be optimized to maximize gas oil recovery and quality while minimizing resid entrainment through the wash zone. The design goal is to achieve two theoretical stages in the wash zone but avoiding wash bed coking is a primary concern so KBC limits the bed height for more severe, heavy crude designs.

Liquid holdup time in the wash bed must be minimized and good liquid wetting of all surfaces must be maintained. For light crude, vacuum column operations where coking is not a design consideration, "Y" type packing can be used in the wash zone. "Y" packing, with surfaces angled at 30° to the horizontal, is more efficient and provides more theoretical stages per height than "X" packing. However, "Y" packing also retains more liquid for a longer time in the packing. For heavy crude deepcut designs where the wash bed is more prone to coke, KBC typically specifies "X" packing with surfaces at 60° to the horizontal to minimize holdup. KBC heavy crude, deepcut designs also specify that the structured "X" packing be supported over open grid and the height ratio of the packing to grid is lower for more severe operations. Although coking is mitigated, fractionation efficiency decreases with more grid in the bottom of the wash bed.

The Hot Wash Oil rate to the top of the wash bed is determined by the depletion ratio through the wash zone and the target overflash or wetting rate at the bottom of the bed for the specific packing, grid and internals. If the overflash slop wax is not measured but flows through internal pipes into the bottom of the column, it is difficult to maximize cutpoint because the overflash must be determined by simulation. Even when the overflash slop wax is measured, accurate readings are questionable with conventional flow meters. Although it is beyond the scope of this discussion, the design of the vacuum column internals and the primary goal of mitigating coking and plugging of the wash bed can limit cutpoint and gas oil recovery in existing units.

Overflash slop wax disposition

Some vacuum unit designs recycle the overflash slop wax back to the inlet of the vacuum charge heater for the purpose of recovering gas oil from the overflash. KBC studies show that this is not as efficient as steam stripping the gas oil from the overflash in a wet vacuum unit design with a bottoms stripper. Overflash typically contains approximately 40-60% entrained vacuum resid and should be sent to the stripper. If the unit is charge heater limited, recycling the overflash to the heater limits vacuum resid feed. KBC designs with bottoms stripping steam withdraw the vacuum column overflash to ensure no quenching of the wash zone and pump the overflash back to the vacuum column on the top tray of the bottoms stripper. Most of the cutpoint heavy gas oil is stripped in the bottoms stripper to maximize cutpoint and gas oil lift. The overflash flow is measured with a mass flow meter and the hot wash oil adjusted to maintain the target wetting rate at the bottom of the wash bed.

Dry versus wet

If the existing vacuum unit design is dry with no heater steam and no bottoms stripping steam, then HVGO cutpoint is typically limited by the vacuum heater cracking curve for heavy crudes. KBC experience is that deepcut operation is only attainable in dry vacuum units for relatively light crudes with low asphaltene content. A definitive answer can only be obtained by rigorously simulating the vacuum column and vacuum heater to determine if cracking is an issue at the target cutpoint. In general, KBC studies have shown that dry operations can achieve 1000°F cutpoints but a 1050°F cutpoint is questionable for dry operations and cutpoints approaching 1100°F will be constrained by cracking limitations with a dry operation.

Dry operation is always at lower pressures than an equivalent wet operation. With no heater velocity steam to lower the effective hydrocarbon partial pressure, the column pressure must be reduced to obtain the required vaporization and gas oil lift. However, this has little effect on cracking in the heater since the transfer line pressure drop will increase at higher velocities. Well designed, dry vacuum columns operate at approximately 5 to 8 mm Hg(a) at the overhead of the column and 12 to 15 mm Hg(a) in the flash zone. Optimum pressures for the wet operation are 10 to 15 mm Hg(a) overhead pressures and 30 to 45 mm Hg(a) in the flash zone. However, since cracking is not as much of a concern, wet columns are sometimes operated at elevated pressures to reduce the hydraulic loadings, i.e. C-factors in the wash zone for existing units that are limited by the column diameter.

Vacuum heater coking

KBC analyzes vacuum heater and transfer line performance in one integrated VIS-SIM model. The proprietary simulation of the vacuum heater provides a rigorous tube-by-tube calculation of bulk and film temperatures. Pressure drop calculations check the transfer line critical and sonic velocities which can contribute to entrainment in the vacuum column. For each convection tube, crossover, radiant tube, each change in the outlet piping diameter, headers and transfer line, VIS-SIM reports accurate process pressures, temperatures, oil film temperatures, tube metal temperatures, fluid velocities, critical velocities, sonic velocities and the percent vaporization.

What is unique is that VIS-SIM can generate KBC proprietary cracking curves for a range of heater conditions. Residence time versus film temperature plots facilitate the analysis of each point in the heater, outlet piping and transfer line to ensure operation below the cracking curve. The KBC proprietary correlations correlate the occurrence of cracking to film temperature and the residence time at each point in the system. For two-phase flow regimes, VIS-SIM generates Baker charts for the horizontal and vertical flow in the tubes, headers and transfer line to ensure the design will operate outside undesirable flow regions. VIS-SIM ensures that neither the heater nor transfer line will operate in stratified, slug flow or separate phase, flow regimes.

Vacuum heater velocity steam

If cracking becomes a concern, the process conditions must be modified and this is typically accomplished with the addition of heater coil (velocity) steam which is why dry operations will limit heater temperatures and cutpoint. Many KBC studies indicate that deepcut operation is not attainable for heavy crude slates in a dry system. However, the design studies also show that in a wet operation, bottoms stripping steam is much more efficient at increasing gas oil lift and recovery from vacuum resid. Standard vacuum heater designs target 2.0-2.5 wt% steam but a rigorous VIS-SIM tube-by-tube cracking analysis facilities shifting coil steam to bottoms stripping steam which can increase gas oil lift and cutpoint at the same ejector process water load.

Vacuum transfer line limits

The vacuum transfer line has been an area of focus in the industry. There have been many articles about sonic velocity or descriptions of simulation techniques using a series of flashes to model phase separation with diagrams showing separate vapor and liquid phases flowing through the transfer line. KBC has designed transfer line systems using VIS-SIM for diameters ranging from 56 to 90 inches and checked some piping layouts with Computational Fluid Dynamics (CFD) models. Phase separation is not an issue and the CFD simulations show homogenous phase distribution through the cross section of even large diameter transfer lines. The models show the vapor and liquid are well-dispersed because transfer line velocities are typically 200-300 fps and the velocity increases from the heater outlets as the pressure drops.

A Yokogawa Company

Older vacuum unit designs incorporated transfer lines with a telescoping configuration that were sized in an attempt to maintain the critical velocity below 80% of the sonic velocity at any point. While this may have seemed like a reasonable design consideration, KBC designs studies have shown that it is difficult to implement. Strictly adhering to these guidelines will result in a long, tortuous piping layout. The telescoping transfer line is mechanically difficult to construct and the design limits operating flexibility. This approach also increases pressure drop.

KBC design experience is that the best approach is to size the transfer line to operate well below sonic for some distance before the flash zone inlet and to take the sonic drop closer to the heater outlets, typically at the risers from the individual passes. The KBC design allows a normal flow pattern to be re-established upstream of the flash zone inlet distributor. Any fine liquid spray generated by the sonic pressure drop will disperse before entering the column. KBC is more concerned about the ability of the inlet distributor to coalesce the resid and avoid entrainment through the wash zone. The KBC design discharges the heater outlet lines horizontally at 45° into one large diameter transfer line which is piped directly into the flash zone inlet distributor to minimize the transfer line length. CFD studies show no phase separation in the large diameter transfer line and minimal liquid at the heater outlets.

KBC evaluations of existing vacuum units show that the transfer line can limit either the heater or the column operation. An undersized transfer line will increase the pressure at the heater. This in turn will result in higher film temperatures at constant flash zone conditions and drive the heater toward a cracking regime. An undersized transfer line that results in sonic flow close to the tower may result in poor fractionation between the heaviest sidestream and the bottoms. The use of an equilibrium model for vacuum unit transfer line design has been the traditional method for modeling the vapor-liquid fluid properties across the piping into the vacuum column. There have been claims that a more accurate model of the transfer line can be achieved using a non-equilibrium approach. This has created some debate as to how new transfer lines should be designed, especially as it impacts the vacuum column design and the maximum cutpoint. Computational fluid dynamics, Petro-SIM process simulations, design tools and VIS-SIM plus operating data for existing units have provided additional insight to clarify design parameters and ensure a robust, yet cost-effective, vacuum heater, transfer line and column design.

WHAT IS THE OPTIMUM GAS OIL CUTPOINT?
So what is the optimum cutpoint? There are multiple definitions of cutpoint used in the refining industry and some facilities believe that if there is minimal overlap in the 5% VTB - 95% HVGO then this is the TBP Cutpoint or HVGO Final Cutpoint (FCP) and VTB Initial Cutpoint

(ICP). However, the disposition of the Slop Wax Overflash must be considered and the true cutpoint is based on the lowest sidedraw and the vacuum column bottoms. Assuming the disposition of the slop wax does not affect the cutpoint calculation and that HVGO is the lowest sidedraw, calculating the cutpoint based on operating data is still difficult without a model. KBC uses Petro-SIM to back-blend the actual operating yields and lab results and then recut the synthesized, mass balanced feed to those yields to calculate the product cutpoints.

The KBC definition of cutpoint is based on cutting the TBP curve for the vacuum feed at the volumetric fractions of the products. The HVGO cutpoint is the final vapor temperature of the vacuum column feed on a TBP basis when HVGO and all lighter products have been removed. TBP initial cutpoints and final cutpoints are accurately calculated for crude and vacuum unit products in the Petro-SIM column simulation but are not easily obtained from operating data. Since the vacuum bottoms initial cutpoint is based on the distillation of the vacuum unit feed, the lowest sidedraw and the vacuum bottoms, full range distillations may be required for not only HVGO but also the Atmospheric Resid and Slop Wax Overflash plus a front end distillation to cracking for the Vacuum Resid. Then there is the question of the accuracy of lab distillations on resids that contain asphalt. If the lab analysis is a High Temperature Simulated Distillation does this correlate to a D-1160 Vacuum Distillation and D-2892 True Boiling Point (TBP) distillation.

The vacuum unit must mass and volume balance to determine the volume yields of the products based on the vacuum column charge to calculate the product cutpoints. Many vacuum units receive hot atmospheric resid directly from the crude column bottoms. Without mass flow meters, what is the accuracy of volume correction factor from 650°F to 60°F.? Furthermore, the viscosity correction for resids can be as high as 6 to 8%.

In summary, the determination of the vacuum resid initial cutpoint on a TBP basis is challenging when analyzing plant data. High TBP cutpoints are frequently quoted as the optimum deepcut operation but rigorous modeling of the units have often shown the actual cutpoint is different. Many KBC refinery-wide Petro-SIM flowsheet studies have shown that high cutpoint operations are not profitable for some refinery configurations, vacuum unit designs and the downstream conversion units. KBC experience is that deepcut studies cannot be conducted without a refinery-wide Petro-SIM flowsheet simulation.

KBC Advanced Technologies Inc - Regional Headquarters
15021 Katy Freeway, Suite 600, Houston, TX 77094, USA

T: +1 281 293 8200 / F: +1 281 616 0900 / E: info@kbc.global / www.kbc.global

513958 Building Flexibility into a Refinery with a Delayed Coking Unit

Abstract:

Very few refineries process a single crude, or for that matter, the crude that their configuration was designed to process. Often, the refinery must cope with a multitude of opportunity crudes as they try to maximize their margins while keeping within the limits of their configuration. Delayed coking provides a refinery operator with unparalleled flexibility to process the vacuum residue from widely varying crude oils. Unlike catalytic processes, delayed cokers are largely insensitive to feedstock properties, including the level of contaminants such as sulfur, metals and Conradson Carbon Residue (CCR).

This presentation will summarize the leading residue upgrading technology options, and then examine the effect on the coker resulting from a significant change in crude composition to a refinery. Flexibility can make the difference in todayâ€™s competitive market, and a Delayed Coker has the in-built flexibility to process whatever crude oils the traders are likely to buy for the refinery.

513495 Smart Systems for the Process Industries

Abstract:

ABSTRACT for AIChE 2018 Spring Meeting

Smart Systems for the Process Industries

Samvith Rao and James C Cross III

The MathWorks, Inc.

samvith.rao@mathworks.com

508-647-4487

An unprecedented and increasing level of device and computing capability and connectedness is intensifying competition and paving the way for new business opportunities. The challenge for existing companies and new ventures alike is to: determine what information is important; decide how to collect and process it; and conceive, develop, and offer new products and services that deliver value from it. This is the aspiration of the Industrial Internet of Things (IIoT), and Industry 4.0.

Specifically, the proliferation of inexpensive sensors, wireless communications, big data and cloud infrastructure, machine and deep learning algorithms, and ever-faster high-performance computing platforms is ushering in a new era of intelligent and autonomous systems, including self-driving cars, smart industrial systems, and automated warehouses. Moreover, the confluence of these technologies has given rise to a new type of industrial asset – a virtual, real-time replica of a companion physical system, the so-called *digital twin* – that can be leveraged for a wide range of operational benefits, including process optimization, predictive maintenance, fault detection, model-based control, and dynamic pricing.

In this talk we describe the individual enabling technologies, present examples of how they are being developed or used commercially, and offer perspectives on their adoption and potential within the process sector.

512516 Smart Damper Design Provides Better Control of Fired Heaters

Abstract:

Fired heaters are major consumers of energy in the Refining and Petrochemical Industries. In most of the natural draft fired heaters, one of the important parameters that is controlled by the operators is the furnace draft using stack dampers. Historically stack damper quality in the industry is very poor. The current design of the stack dampers is inadequate to control the draft effectively. As a result most of the stack dampers are left fully open in the industry. The average draft in the fired heaters is maintained at almost 3-4 times the value recommended causing a lot of tramp air in the system.

In this paper; we are going to talk about how our patent pending Smart Stack Dampers can ensure proper draft control in fired heaters. We have made several improvements in the design and construction of the dampers. Smart dampers are designed to be very reliable and operator friendly. They provide a much better control of the draft. Smart dampers can be configured for automatic or manual draft control of the heater

513860 Application Centralized Information Technology in Process Design Package Development

Abstract:

Smart Manufacturing solution is built on information technology, researchersâ€™ ingenuity to create innovative product, continuously analyzing market and manufacturing data to discover the market needs and signal problems impacting your costs, quality, and production of the existing technology. Centralized database technology with advanced simulation are keys to quickly delivery of the new products reliably, safely, and environmental friendly.

This presentation provide an introduction for applying centralized information technology in process design package development (1) evaluating new innovative technology for its reliability, product profitability, global market potential, sustainability and technology; (2) Aligning proven and conventional technology to form an integrated commercial process based on commercial database; (3) develop the process design package and expand the process globally in a short timeframe.

512754 Advances in Utilizing Data to Improve Operations

Abstract:

Big data and Industrial Internet of Things (IIoT) are relevant topics of discussion in many commercial and industrial applications, and it also applies to the operation of refineries and petrochemical plants using analyzed data to improve operations. Online data from sensors, whether wired or wireless, can be analyzed and used to provide plant operators with the actionable information to make timely decisions for improving asset reliability and efficient operations.

Properly trained operations are invaluable in this process, as they not only act on information from analyzed data, but can also make decisions on their own based on sensor data.

Online data from sensors has been available for decades, but a transformation is now taking place due to the low cost and quick installation time for WirelessHART® (an international industry standard IEC 62591 since 2010) sensors when compared to their wired counterparts. The term pervasive sensing describes the application of wireless sensors to areas which were formerly too expensive to monitor with wired instruments, leading to the creation of big data.

Abnormal operation can quickly be identified and corrected with properly trained staff to act on this new timely analyzed information. Big data and IIoT will continue to have a role with shortages of skilled resources, thus having the staff be more efficient and effective at their role with additional information and insight into operations.

This presentation will focus on the applications where big data and IIoT have improved operations. Although the applications presented will come from oil refineries, the solutions can be applied to other continuous processing facilities.

Adaptive anomaly detection and its application to online asset monitoring

Dr Tim Butters, CMath MIMA

Head of R&D – Sabisu, United Kingdom
tim.butters@sabisu.co

Abstract

The automatic detection of faults in assets is a challenging problem, with complexities arising due to the constant changes in conditions seen in industrial operations. Here we present a method of time series anomaly identification that successfully identifies this incipient behaviour whilst adapting to the changing plant conditions. The system was applied to a hyper-compressor and used to detect packing ring failures with a high level of accuracy.

1 Introduction

Industrial plants are comprised of a large number of assets each with associated process data, much of which is digitally historised at a high frequency. This process data can be monitored either manually by engineers or automatically by software systems to identify incipient issues that need addressing. Software systems are preferable as they can monitor continuously, however, with plants constantly changing operating modes and the influences of external factors it is extremely difficult to develop a fault detection system that operators well in the general case.

Instead, algorithms that tackle specific asset failure modes are developed. Although the effectiveness of such systems can be high, the lack of horizontal applicability severely limits their use. In this work we present a method for adaptive anomaly detection in time series data that can be effectively used for asset fault detection whilst accounting for changing plant conditions.

Asset failure modes are often complex. Even with multiple examples of the same failure type there can be significant differences between the process data for each case. This, coupled with the limited number of examples of each failure mode available in an industrial setting can make data-driven modelling of asset failure using typical machine learning techniques very difficult. To overcome these issues, we have developed a method of data-driven anomaly detection that can effectively identify anomalous time series behaviour based on a small portion of past "normal" operation. This removes the reliance on large datasets associated with typical supervised or unsupervised machine learning solutions.

This anomaly detection system was used as the basis for an online monitoring solution for a two stage hyper-compressor. The specific problem being addressed is the identification and tracking of packing ring failures for the compressor to eliminate the corresponding unplanned down-time.

2 Anomaly Detection

There is no formal definition of an anomaly in the general case of time series data. For practical purposes we define an anomaly as a short period of data that does not match the behaviour of its recent past. For most asset faults the corresponding features seen in the process data fit this description, and are detectable through methods specifically designed with this definition in mind.

2.1 Algorithms

In this system several anomaly detection algorithms are used to provide flexibility in the types of abnormal behaviour that can be found. This allows the system to be uniformly deployed over different assets with minimal configuration. Below is a list of the algorithms we have implemented and tested. All of these algorithms build a statistical model of the time series, triggering anomalies when the data fails to fit this model. The data model is continually updated in real-time, which assures that when the normal behaviour of the asset changes the algorithms will adapt to this new state automatically.

AdWin

AdWin is a change-point detection algorithm developed by Bifet and Gavaldà [1]. It searches for points in the data that separate periods of behaviour based on the mean value within an expanding window. It is therefore very effective at locating changes in average value, but it is insensitive to changes in the variance of the signal.

StatWin

To capture changes in variance we developed an anomaly detection algorithm called StatWin. StatWin detects when the most recent data point lies outside of a statistical envelope defined by previous behaviour within a moving time window.

KMeans

A k-means clustering algorithm is used to model the data, and then the most recent data value is compared to this clustering to determine if it is sufficiently "close" to the cluster centres formed within a fixed window. If it is not, the value is considered anomalous.

This algorithm is based on an approach described by Hill and Minsker [2]. It effectively detects changes in both average value and variance, but rather than relying on a continuous window of allowable data it uses the cluster centroids to define regions of acceptable values. This means that this algorithm may recognise anomalies that lie within the range of previously seen data. For example, given a signal comprising entirely of the numbers 0 and 100, this algorithm would indicate a value of 50 as anomalous, as it is not sufficiently "close" to the previous values.

2.1.1 BCPD

On-line Bayesian change-point detection (BCPD) builds a statistical model of the data and marks the points at which the underlying data distribution changes significantly [3, 4]. This approach typically provides a good general purpose algorithm that is directly applicable to any time series without any changes in configuration.

2.2 Data Windowing

Data windowing provides an efficient method of dealing with continuous data in real-time. All but one of the anomaly detection algorithms, AdWin, use a fixed size moving window. AdWin is designed around the concept of a dynamically expanding window that shrinks when changes are detected to exclude the old "different" data. This means that the window expands gradually as new points are added, and then shrinks instantaneously to remove the oldest data which does not fit the current behaviour.

KMeans and StatWin require a fixed size window to be populated before anomalies can be detected. This is because accurate statistical measurements (in StatWin's case) and k-means clustering (in KMeans' case) cannot be performed without a sufficient number of data points. Once "full", new data points are added to the end of the window and the oldest point is removed from the start to maintain a constant window size.

To ensure numerical stability the on-line Bayesian change point detection algorithm also uses a fixed size window, although it does not have the limitation of requiring this window to be fully populated before anomalies can be detected. This change detection algorithm calculates the probability that each data point being analysed is a contiguous block of similar behaviour. Smaller and smaller weights are given to points more distant from the latest value.

2.3 Quorum Voting

In certain circumstances it may be appropriate to use a single anomaly detection algorithm. However, it has been found that combining the outputs of several algorithms minimises the number of false positives detected by the system. To combine the results of the algorithms a quorum voting system is employed. This means that an anomaly is only reported if the majority of the algorithms agree on the result. To account for the difference in techniques between the algorithms the results do not have to align exactly in time, but within an automatically determined timespan.

The quorum timespan can range from a few minutes to hours depending on the frequency of the underlying sensor signals and the behaviour of the asset. This accounts for the

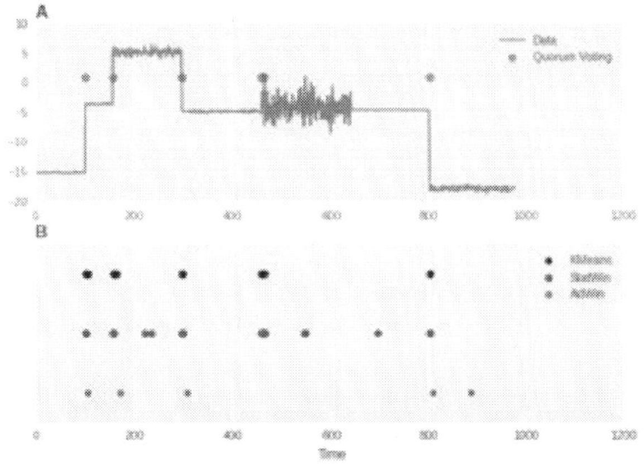

Figure 1: Anomaly detection on a test signal using the KMeans, AdWin and StatWin algorithms. The results from the individual algorithms are shown on panel B, with the signal and quorum voting result shown in panel A. From this image it is clear that the quorum voting only flags an anomaly if two or more of the three algorithms agree.

different ways in which the above algorithms detect anomalies, ensuring that the sensitivity is maintained but the false positive rate is minimised.

An example of anomaly detection is shown in figure 1. The three algorithms KMeans, AdWin, and StatWin are combined to produce the quorum voting results shown as red dots in panel A. The detection results of the individual algorithms are shown in panel B.

3 Case Study

A two-stage hyper compressor was identified as a key asset in a customers process. This hyper-compressor has 20 cylinders, each cylinder has a packing box containing 5 packing rings that are responsible for the sealing of the system. The packing boxes must be replaced periodically as over time the rings degrade and fail. If the packing boxes are not changed in time and all the rings fail, an automatic plant shutdown is triggered. This unplanned down-time lasts for approximately 2 days and costs in the region of £500,000 each time.

The packing boxes themselves are sealed and cannot be inspected, but there is a limited amount of process data available for the asset: total leak gas for the whole compressor, plunger temperature for each cylinder, and x-runout and y-runout (measures of displacement) for each cylinder. The challenge was to create a system that could monitor this process data to determine when a ring had failed, and in which of the 20 packing boxes the failure had occurred. With this information it would be possible to track the status of each packing box and schedule replacements accordingly, avoiding both premature replacement and unplanned outages due to total packing box failures.

To achieve this the anomaly detection system presented above was applied using a two-stage approach. It was found that the total leak gas measurement showed a clear spiking signature when a packing ring failed. This signature was accurately identified by the anomaly detection system with a very low false positive rate.

To identify the specific packing box containing the failed ring the anomaly detection system was then run over the cylinder-specific measurements for each cylinder (60 time series in total). Correlation analysis was then used to find any anomalous behaviour in these signals that corresponds to the anomaly found in the total leak gas. This successfully identified the packing box containing the failed ring.

The overview screen of the system is shown in figures 2, it shows a schematic of the compressor with a colour-coded health indication for each cylinder. The health is determined by the number of anomalies attributed to the cylinder's packing box, which correlates to the number of rings remaining. It is possible to have more anomalies detected than rings in the box as a ring can fail in several stages over a period of a few hours.

An example ring failure is shown in figure 3. The point at which the anomaly was detected is marked with a small Sabisu logo in the centre of the plot. The leak-gas, shown in blue on the chart, gives a clear indication of a failure. The corresponding activity in the cylinder specific measurements (yellow, green and red lines) indicates that this is the cylinder that has experienced the failure. Axis labels have been removed inline with a confidentiality agreement.

This compressor monitoring approach was found to be a highly effective method for identifying failed packing rings. As well as successfully identifying all the historical ring failures including the affected packing box, it has also found several confirmed packing ring failures in the live environment since it has been deployed to the customer site. It is now used to determine the packing box replacement schedule. Since its deployment in October 2017 it is estimated to have saved the customer £800,000 in lost production.

4 Discussion

4.1 Anomaly Detection

Here we have presented a novel method of time series anomaly detection. This method comprises of several time series analysis algorithms that build a statistical picture of the data and highlight points that do not match the picture built using the recent data history.

This method is an effective strategy for detecting anomalies whilst adapting to changing plant conditions because the data model is being constantly updated. This allows the system to implicitly learn the new "normal" state without any user intervention.

The presented algorithms are neither processor nor memory intensive, meaning that large datasets can be processed quickly and efficiently. As the algorithms build an ever changing view of the data there is no need to store entire datasets in memory for processing, allowing them to be used in an online real-time setting, and removes any limits on the size of the data that could be processed using this technique.

There is a small amount of configuration for the StatWin and KMeans algorithms, setting the threshold at which an anomaly will be flagged. These parameters are easily configured using periods of known anomalous behaviour and requires no special training. Default values for these parameters are provided

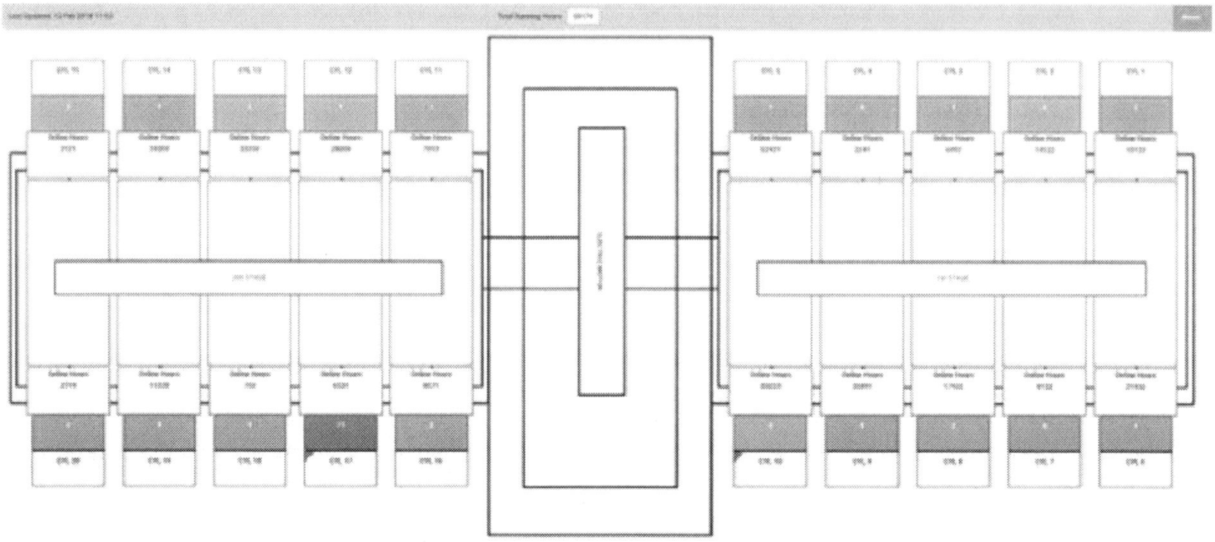

Figure 2: The overview page of the compressor monitoring system gives a clear view of asset health. Each cylinder has a colour-coded status, with the number of anomalies detected displayed for extra detail.

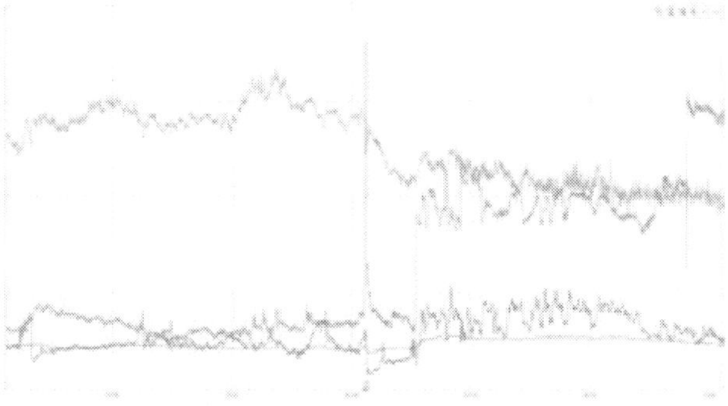

Figure 3: Data view from the Sabisu platform. This shows the compressor sensor data with a detected anomaly marked with the small Sabisu logo. The leak-gas measure shown in blue gives a clear indication of a fault, with corresponding activity in the other signals identifying this cylinder as the one experiencing the ring failure.

that have been suitable estimations in the majority of deployments of this software. It is also possible to alter the size of the sliding window used to define the data model, however, by default it is automatically determined based on the frequency of the data which has been a suitable method in all of our use cases to-date.

4.2 Industrial Case Study

As can be seen from figures 2 – 3, the presented system successfully identifies asset failures and can be used to create asset monitoring solutions. The ring failures are detected before the plant engineers had identified the fault. As well as the large number of time series to be monitored this is also partly due to the nature of this particular type of failure. As the signals quickly recover to normal conditions, if only a short section of data is assessed (*e.g.* one day) it is easy to miss a failure that happened just a few hours earlier.

The number of anomalies detected by any fault identification system is an important aspect in its "trustworthiness". As this system is designed to detect anomalous behaviour across different asset types with minimal configuration it is inevitable that there will be some false positives or missed anomalies. The preferred state is a small number of false positives with no missed true faults rather than a more conservative state in which some faults are missed. However, if the system needs to be configured such that false positives are eliminated at the potential cost of missed anomalies this is easily configured through the detection threshold parameters.

The false positive rate depends on the variability of the data being analysed. The rate in the example of compressor monitoring is extremely low, with one presented by the system over a period of approximately 6 months. In a different application of heat exchanger tube leak detection the system had a higher false positive rate, but it was still considerably lower than the acceptable rate specified by the customer in question.

5 Conclusions

An effective anomaly detection system has been presented, capable of detecting various types of anomalous behaviour across a large range of industrial assets. The system requires very little configuration, and quickly adapts to changing behaviour to allow its continued use for long periods of time without user intervention. The utilisation of quorum voting minimises the false positive rate of the system whilst retaining the sensitivity of the individual algorithms.

It has been demonstrated that the system successfully identifies packing ring failures in a two stage hyper-compressor, providing crucial information to engineers that would otherwise not be available. The system correctly identifies all historical ring failures, and since being online has saved approximately £800,000 in lost production since October 2017.

References

[1] A. Bifet, R. Gavaldà. "Learning from time-changing data with adaptive windowing", *In SIAM International Conference on Data Mining*, (2007).

[2] D. J. Hill, B. S. Minsker. "Anomaly detection in streaming environmental sensor data: A data driven approach", *Environmental Modelling & Software*, **25(9)**, pp. 1014–1022, (2010).

[3] R. Adams, D. MacKay. "Bayesian online changepoint detection", *University of Cambridge Technical Report*, (2007).

[4] P. Fearnhead, Z. Liu. "On-line inference for multiple changepoint problems", *Journal of the Royal Statistical Society: Series B (Statistical Methodology)*, **69**, pp. 589–605, (2007).

Dynamic Data Analysis™ of Large Scale Data to Monitor Fouling in Heat Exchanger Networks

Francesco Coletti[1,2*], Emilio Diaz-Bejarano[1], and Sandro Macchietto[1,3]

[1]*Hexxcell Ltd., Innovation Hub, Imperial College - White City Campus, 80 Wood Lane, London W12 0BZ, UK.*

[2]*College of Engineering, Design and Physical Sciences, Brunel University London, Uxbridge, UB8 3PH, UK*

[3]*Department of Chemical Engineering, Imperial College London, London SW7 2AZ, UK*

* f.coletti@hexxcell.com

Abstract

Data collected from heat exchangers networks in the field are typically used to monitor the thermal performance of the individual units. However, the information extracted from the data is limited in quantity and quality by the simplified models typically used in practice. Key decisions such as cleaning of heat exchangers rely on the calculation of derived quantities such as the fouling resistance which lumps together a number of factors contributing to fouling. This approach has been severely criticised in the past by various authors but it is still widely used in the industrial practice [1]. In this paper it is shown that a significantly larger amount of information and insights can be extracted from the same measurements by using rigorous models and a flexible framework. It is shown how this approach leads to a much deeper analysis of the status of the network which, in turn, helps with diagnosis and troubleshooting. An industrial case study is presented to illustrate the benefits.

Introduction

With the reduction in cost of sensors technology and IT infrastructure, the process industry is experiencing an unprecedented explosion in the amount of information being collected from each piece of equipment in each unit operation on every site. However, these data are often not exploited, particularly in the analysis of fouling in heat exchanger networks.

This paper presents an in-depth methodology that exploits existing measurements and data to monitor fouling in heat exchanger networks. The methodology, powered by Hexxcell Studio™, consists of a detailed dynamic analysis that uses historical plant data in conjunction with advanced mathematical models [2–4] to provide deep insight into the thermo-hydraulic and fouling behaviour of heat exchangers. By using this systematic analysis, it is possible to assess the quality of the measurement used, the thermo-hydraulic performance of individual heat exchangers and the entire network, the economic impact of fouling and the efficiency of cleaning actions. The analysis also allows identifying opportunities for improvement in the network (e.g. different flow split operation, cleaning strategy etc), diagnose causes of fouling and detect changes in behaviour (e.g. inorganic breakthrough) and assess the effectiveness of mitigation strategies implemented such as antifoulants, revamps etc. Combined with blending information, it is also possible to unveil key correlations between operating parameters, crude slates and fouling behaviour. An industrial case study illustrating the above benefits is presented.

Method

The dynamic analysis of the data (Dynamic Data Analysis™) is performed by using an advanced heat exchanger model [2, 3] whereby the evolution of the characteristics of the deposit is calculated directly from the plant data as a function of measured inlet conditions of temperature and flowrate, heat duty calculated from primary measurements and (where available) pressure drop measurements (a full example of this is provided in [6]). In the absence of the pressure drop measurements (which is a very typical situation in oil refineries), the deposit thickness can be calculated by fixing the deposit thermal-conductivity, allowing the calculation of hydraulic variables under reduced flow area conditions due to the presence of the layer: fluid velocity, shear stress, pressure drops, etc.

Case Study

The case study focuses on the hot end of the preheat train of a mid-size oil refinery, comprising of a total of five heat exchangers. The data analysis was performed on a set of plant measurements (temperatures, flowrates) for an operating period of ca. 1200 days.

Results

The analysis provides a comprehensive insights into the evolution of fouling, the performance of the exchanger, operating conditions and dimensionless numbers of interest on both tube and shell sides. Here, we focus on some of the most relevant findings for this case study.

First, the evolution of the fouling behaviour was obtained for the entire period in a seamless simulation. For the sake of comparison, the overall heat transfer fouling resistance for three of the units is shown in Figure 1. The analysis helped quantify the efficiency of intermediate cleaning actions carried out at the refinery. Cleaning actions are indicated in Figure 1 with labels M for mechanical and C for chemical and are shown as sudden drops in the fouling resistance. In this case the analysis also helped identify differences in fouling behaviour between various units. The three units in Figure 1 share very similar tube-side design and operating conditions. However, the fouling resistance in E04 is approximately twice that in E02AB and E05AB indicating a possible shell-side fouling build-up in E04 which does not affect other units [7].

By considering tube-side fouling in all units, and a 50/50% distribution of fouling between tube and shell sides in E04, the analysis provided the thermo-hydraulic evolution of the performance of the various units shown in Figure 2a. The thermo-hydraulic plot (TH plot) displays the ratio of the fouled-to-clean heat duty against the ratio of the fouled-to-clean tube-side pressure drop.

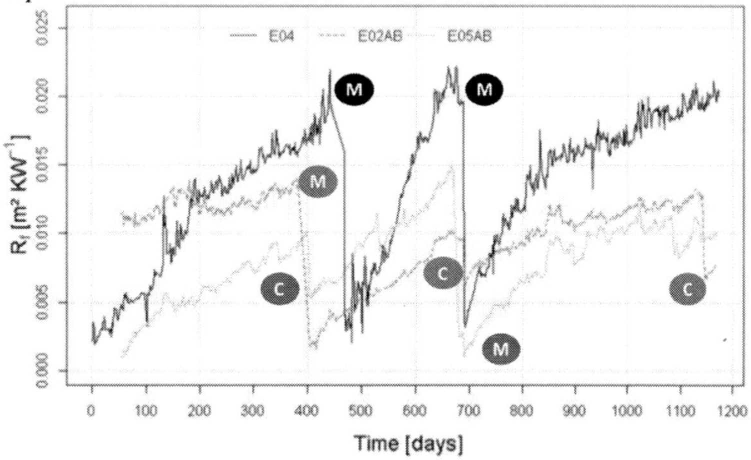

Figure 1. Total fouling resistance in E02AB, E04 and E05AB. M and C indicate mechanical and chemical cleaning, respectively (adapted from [7]).

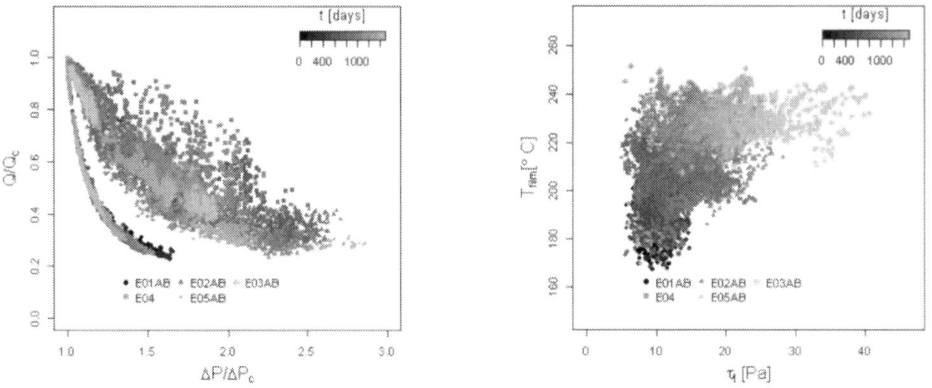

(a) (b)

Figure 2. Tube-side TH-plot (a) and tube-side film temperature against shear stress (b) under fouled conditions over a period of ca. 1200 days.

As fouling builds up, heat transfer gradually degrades and the cross-sectional flow area is restricted, which decreases the heat duty and increases the pressure drop. Simultaneously, the analysis provides the operating conditions at each point of the exchanger under fouled conditions. Figure 2b shows the tube-side film temperature against the tube-side shear stress. The figure shows that these conditions, which are typically used to characterize fouling behaviour, vary significantly during the operation. This variability is result of a combination of process variability and fouling build up. The latter can be visualized as the gradient from darker points (older in time) to paler points (latter in time). Figure 2 also shows that E04, with the fouling distribution selected, indeed operates under similar conditions and performance than E02AB and E05AB.

Finally, an extension of the method, the Fouling Propensity Analysis™, can be used to estimate fouling rates and find correlations between operating conditions and the concentration of the oils processed. This allows identifying oils that are particularly problematic with respect to fouling deposition or, conversely, oils that have a beneficial effect and help reducing the fouling resistance. Figure 3a shows the results of the correlation analysis between fouling rates and the mass fraction of the oils processed in the refinery. The information on oil composition has been modified to protect confidentiality without affecting the significance of the example presented.

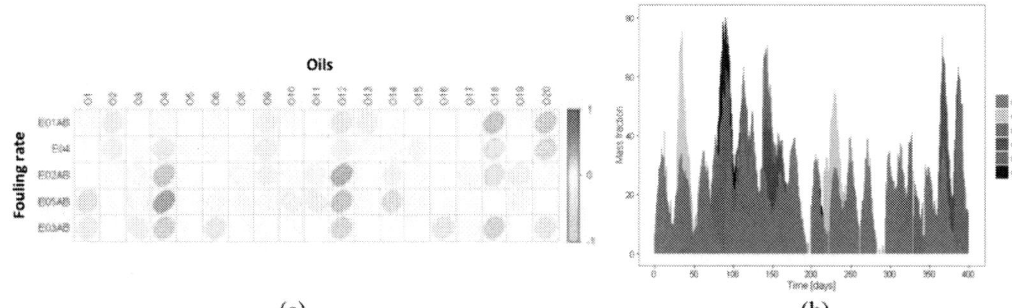

(a) (b)

Figure 3. Result of correlation analysis between fouling rates and mass fraction of the oils processed (a) and processing history of oils identified as problematic (b). Information on oil composition has been modified to protect confidentiality without affecting the significance of the example presented.

Conclusions

An industrial case study has been presented in which Dynamic Data Analysis™ allowed studying the evolution of the fouling behaviour and its impact on the operating conditions inside the tubes (film temperature, shear stress) and the overall thermal and hydraulic performance of the exchangers. The analysis allowed identifying and quantifying the extent of cleaning actions and detecting the different fouling behaviour in one of the units, likely due to severe shell-side fouling. Finally, the fouling rates obtained were cross-correlated to information available on the crude oil slates processed. This permitted identifying oils particularly problematic towards fouling deposition.

References

[1] F. Coletti and G. F. Hewitt, Crude Oil Fouling: Deposit Characterization, Measurements, and Modeling. Boston: Gulf Professional Publishing, 2014.

[2] F. Coletti and S. Macchietto, "A Dynamic, Distributed Model of Shell-and-Tube Heat Exchangers Undergoing Crude Oil Fouling," *Ind. Eng. Chem. Res.*, vol. 50, no. 8, pp. 4515–4533, Apr. 2011.

[3] E. Diaz-Bejarano, F. Coletti, and S. Macchietto, "A new dynamic model of crude oil fouling deposits and its application to the simulation of fouling-cleaning cycles," *AIChE J.*, vol. 62, 1, pp. 90–107, 2016.

[4] E. Diaz-Bejarano, F. Coletti, and S. Macchietto, "Impact of Complex Layering Structures of Organic and Inorganic Foulants on the Thermohydraulic Performance of a Single Heat Exchanger Tube: A Simulation Study," *Ind. Eng. Chem. Res.*, vol. 55, no. 40, pp. 10718–10734, 2016.

[5] E. Diaz-Bejarano, F. Coletti, and S. Macchietto, "Thermo-Hydraulic Analysis of Refinery Heat Exchangers Undergoing Fouling," *AIChE J.*, vol. 63, no. 3, pp. 984–1001, 2017.

[6] E. Diaz-Bejarano, E. Behranvand, F. Coletti, M. R. Mozdianfard, and S. Macchietto, "Organic and Inorganic Fouling in Heat Exchangers – Industrial Case Study: Analysis of Fouling State," *Appl. Energy*, vol. 206, pp. 1250–1266, 2017.

[7] E. Diaz-bejarano, F. Coletti, and S. Macchietto, "Modeling and Prediction of Shell-Side Fouling in Shell-and-Tube Heat Exchangers," *Heat Transf. Eng.*, vol. Accepted f, no. https://doi.org/10.1080/01457632.2018.1446814, 2018.

Optimization of Refinery Preheat Trains: Predictive Maintenance and Operations Improvement

Francesco Coletti[1,2*], Federico Lozano[1], Emilio Diaz-Bejarano[1], and Sandro Macchietto[1,3]

[1]Hexxcell Ltd., Innovation Hub, Imperial College - White City Campus, 80 Wood Lane, London W12 0BZ, UK.

[2]College of Engineering, Design and Physical Sciences, Brunel University London, Uxbridge, UB8 3PH, UK

[3]Department of Chemical Engineering, Imperial College London, London SW7 2AZ, UK

* f.coletti@hexxcell.com

Abstract

Deciding which heat exchanger to clean, when to clean and how to clean in refinery pre-heat trains is a challenging activity that typically relies on operator's experience. In this paper, an algorithm that allow identifying the most economic cleaning schedule for a given refinery configuration and operating conditions is presented. The method relies on an advanced framework that incorporates rigorous heat exchanger models capable of predicting the fouling behaviour of the refinery as a function of configuration of the individual units and the network, process conditions and time. An industrial case study is presented to illustrate the benefits of the approach, showing that significant improvements over current practice can be obtained.

Introduction

The atmospheric crude distillation unit is the largest energy consumer in the entire refinery. Thus the efficiency of the pre-heat train, which recovers heat from the hot fractionation products, plays a pivotal role for profitability and sustainability [1]. Fouling, the deposition of unwanted material on thermal surfaces, hinders the ability of the pre-heat train to recover heat and forces operators to burn more fuel at the furnace leading to increased costs and environmental impact as well as loss of throughput when the firing limit is reached.

In common practice, maintenance decisions, such as which heat exchanger to clean and when, are driven by a limited amount of information (typically the calculation of the fouling resistance) and operators' past experience. In this paper, advanced models [2]–[4] are used to predict fouling in refinery preheat trains and provide actionable information for predictive maintenance and operations improvement. An improved method for the simultaneous optimisation of cleaning schedules and optimal operations of flow splits is presented and its benefits illustrated with an industrial case study.

Method

The method, implemented in Hexxcell Studio™, allows quantifying and optimising the interactions between cleaning actions and flow split operations and provides information on i) which is the best heat exchanger to clean based on economic considerations including energy at the furnace, potential loss in production and cleaning costs; ii) the preferred cleaning method (i.e. chemical or mechanical); iii) a full cleaning schedule to be implemented over the next year of operations; iv) an assessment of different production scenarios, v) the optimal operation of flow split for the given cleaning schedule proposed.

The algorithm evaluates different cleaning and/or control scenarios at decision points distributed over the time horizon to be explored. At each decision point, the algorithm selects the best heat exchanger to clean, among those that satisfy certain pre-defined constraints, and the best split fraction, according to the key performance criteria in a future operational window. The decision times can be pre-defined at a fixed frequency, or triggered by threshold value of the operation of the network. The algorithm can handle inequality constraints that set limits on the economic savings, energy savings, and/or coil inlet temperature (CIT) value and allow assessing different types of cleanings. The definition of the cleaning scenarios can include simultaneous cleanings of different shells in the network and various cleaning types (mechanical, chemical, or both).

Case Study

The case study involves the hot end of the preheat train of a mid-size oil refinery comprising of five heat exchangers of which four are arranged in parallel branches. Fouling in each of the exchangers is predicted with deposition models fit to and validated against extensive sets of plant data. For simplicity and for the sake of comparison among different alternatives the inlet conditions are assumed constant. In the example, a logic based on economic key performance indicators (KPIs) is selected as a criterion to optimise the operation of the train over an operation period of 1500 day corresponding to a run in between major shutdowns. The economic trade-off involves the costs of fuel in the furnace (which increase with fouling), possible losses in production if the furnace reaches its maximum capacity and the cost of cleaning. The flow split optimization is analysed in a range of 30 % to 70 %.

Results

Five scenarios are defined to show all the capabilities of the cleaning scheduling algorithm:
1) Cleaning scheduling with fixed split fraction at 50 %;
2) Flow split control, without cleanings;
3) Simultaneous cleaning scheduling and flow split control.

The analysis of the results, summarized in Table 1, show that the schedule implemented by the refinery (Case A) allowed a significant saving, estimated in \$3.57 MM, compared to the no action case (Case 0). However, it is estimated that the cleaning scheduling with a fixed flow split proposed by the algorithm (Scenario 1) would have saved a further \$0.5 MM. Scenario 2 (only flow split control) provides significant savings of \$2.2 MM compared to the no action case (Case 0). This indicates that the branches are not balanced in terms of energy recovery once fouling start occurring thus the original design that may be appropriate in clean conditions, is not the optimal configuration for real life operations. However, optimising the flow split alone does not provide any benefits compared to Case A, showing that implanting a cleaning is necessary to maintain the profitability of the operation. Finally, the results show that the simultaneous optimisation of cleaning scheduling and flow split (Scenario 3) provides the most benefits to the refinery, quantified in \$2.7 MM savings over Case A and in over 50% additional savings compared to the case where the flow split was kept fixed at the nominal value. This is expected as the interaction between the flow split distribution and the cleaning of a heat exchanger provides more alternatives that satisfy the process constraints, increasing the number of cleanings that are economically advantageous.

Table 1. Results of cleaning scheduling and flow split control defined using Hexxcell studio

Case	Cleaning scheduling	Flow split control	Economics				Cleanings					
			Fuel cost [MM\$]	Cleaning cost [MM\$]	Total cost [MM\$]	Savings w.r.t. A [MM\$]	E01AB	E02AB	E03AB	E04	E05AB	Total
0	No	50%	63.84	0	63.84	-3.57	0	0	0	0	0	0
A	Actual	50%	59.33	0.94	60.27	0	5	3	2	4	3	17
1	Yes	50%	59.12	0.68	59.80	0.47	9	0	0	7	0	16
2	No	Yes	61.55	0	61.55	-1.28	0	0	0	0	0	0
3	Yes	Yes	56.51	1.11	57.62	2.65	14	0	0	13	0	27

Conclusions

An algorithm for the simultaneous optimisation of cleaning schedules and flow split operation, implemented in Hexxcell Studio™, has been described and applied to an industrial case study. The paper highlighted the importance of i) using predictive fouling models in planning pre-heat trains operations, ii) considering simultaneously cleaning actions and flow split control when planning a cleaning schedule. The case study presented allowed to quantify expected benefit of the framework in $2.7 MM savings. by comparing current practice to the model output.

References

[1] F. Coletti and G. F. Hewitt, *Crude Oil Fouling: Deposit Characterization, Measurements, and Modeling*. Boston: Gulf Professional Publishing, 2014.

[2] F. Coletti and S. Macchietto, "A Dynamic, Distributed Model of Shell-and-Tube Heat Exchangers Undergoing Crude Oil Fouling," *Ind. Eng. Chem. Res.*, vol. 50, no. 8, pp. 4515–4533, Apr. 2011.

[3] E. Diaz-Bejarano, F. Coletti, and S. Macchietto, "A new dynamic model of crude oil fouling deposits and its application to the simulation of fouling-cleaning cycles," *AIChE J.*, vol. 62, no. 1, pp. 90–107, 2016.

[4] E. Diaz-Bejarano, F. Coletti, and S. Macchietto, "Impact of Complex Layering Structures of Organic and Inorganic Foulants on the Thermohydraulic Performance of a Single Heat Exchanger Tube: A Simulation Study," *Ind. Eng. Chem. Res.*, vol. 55, no. 40, pp. 10718–10734, 2016.

[5] E. Diaz-Bejarano, F. Coletti, and S. Macchietto, "Thermo-Hydraulic Analysis of Refinery Heat Exchangers Undergoing Fouling," *AIChE J.*, vol. 63, no. 3, pp. 984–1001, 2017.

[6] E. Diaz-bejarano, F. Coletti, and S. Macchietto, "Modeling and Prediction of Shell-Side Fouling in Shell-and-Tube Heat Exchangers," *Heat Transf. Eng.*, vol. Accepted f, no. https://doi.org/10.1080/01457632.2018.1446814, 2018.

513704 Refinery Preheat Train Fouling Management for Enhanced Energy Efficiency

Abstract:

Heat exchanger fouling in modern refineries continues to be a major issue for both energy efficiency and environmental impact. Fouling is most acute in crude distillation units (CDUs). It reduces preheat temperatures, necessitating higher heater duties which incur increased fuel consumption. This in turn leads to increased fuel costs and higher emissions. It also can reduce throughput and lead to reliability and maintenance issues.

This paper describes a method for managing fouling to minimize its effect on cost, the environment, reliability and maintenance. The approach combines advanced process modelling and optimization techniques with the latest data visualization techniques in an Operational Excellence (OE) platform.

A key development is the use of equation-oriented process modelling tools to model network performance through time. By predicting future performance over an operating window, for example up to the next turnaround, overall costs and environmental impact can be calculated and informed decisions made. The approach results in heat exchanger networks which achieve a practical balance between thermal efficiency and operability, with all fouling effects presented to operators both as costs and environmental impact. An added advantage is that when implemented online the tool can immediately detect anomalies in operation that give rise to abnormal rates of fouling.

Another key feature of the approach is the visualization of results. The technology is deployed via various dashboards, depending on the userâ€™s role. One dashboard is used by engineering teams to study â€˜what-ifsâ€™, another by operations staff to ensure day-to-day performance is optimal, and another by the maintenance department to plan maintenance schedules which tie in to cost and environmental objectives. Current plant data and operational status are displayed in real time.

A real-life case study is presented showing how the technology is used to predict the effects of temperature and flow on fouling, and also the effects of more significant process changes such as throughput increase, bypassing exchangers and the effects of removing heat exchangers. Associated cost and environmental benefits are presented.

513172 Optimising Productivity and Energy Consumption in Crude Oil Distillation Units with Investment-Free Solutions

Abstract:

Crude oil distillation is central to petroleum refining. Crude oil distillation units typically consist of various complex columns (prefractionator or preflash, atmospheric and vacuum distillation columns) and their interlinked heat exchanger network. Distillation units are characterised by high product flow rates and high energy consumption, which means that their operation has serious implications with respect to product revenue and operating costs. Moreover, distillation units need to adapt to different operating scenarios, such as changes in throughput, crude and product quality as well as different prices of utilities, crude oil and distillation products. For these reasons, it becomes essential to develop and implement approaches that maximise productivity and minimise energy consumption in these ever-changing operating scenarios.

This work presents a new approach to optimise crude oil distillation units. Operational changes (e.g. on furnace outlet temperatures, flow rates of pump-arounds, distillation products and stripping steam) are systematically identified to increase net profit and to improve the unitâ€™s flexibility in order to process different types of crude oils. The main feature of this approach is the simultaneous consideration of the distillation process and heat exchanger network during optimisation. This feature is facilitated by the development of surrogate models using artificial neural networks to represent the distillation process. Exploiting the synergy between the distillation process and heat exchanger network delivers more practicable and cost-effective solutions than the conventional approach of considering the distillation unit or heat exchanger network individually.

The proposed methodology consists of three main steps: the analysis, modelling and optimisation steps. In the analysis step, data from plant measurements and reconciled simulations are analysed to define representative operating scenarios and to identify underlying trends that can be used to guide the optimisation (e.g. identify the operational variables that dominate the process economics, define the most critical system limitations). In the modelling stage, comprehensive and accurate models are created using artificial neural network models. These models are developed to represent the distillation process using data from rigorous simulations; while mass and energy balances are used to represent the heat exchanger network. In the optimisation step, the models and the understanding from data analysis are implemented in the optimisation framework to find investment-free solutions that improve the performance of the unit.

The approach has been implemented in a complex Chinese refinery consisting of two prefractionators, two atmospheric columns, one vacuum column and a heat exchanger network. Results verified on-site indicate an increase in profits of more than $ 3 million USD per year mainly by increasing the yield of jet product by 1 %, improving the separation gap between jet and diesel products by about 36°C, and increasing the inlet temperature to the atmospheric furnaces by approximately 1°C.

513624 Sorbent-Based Desulfurization Process for Removing Organic Sulfur Compounds

Abstract:

Gas treatment in oil and gas fields, refineries and chemical facilities is getting more complex due to stringent emissions requirement. While the conventional acid gas removal systems based on amine scrubbing and caustic treatment are highly effective in removing hydrogen sulfide (H_2S) and simple mercaptans, these technologies are not highly effective for complex organo-sulfur compounds such as complex mercaptans (e.g., tert-butyl mercaptan), sulfides (dimethyl sulfide, carbonyl sulfide), thiophenes (alkyl thiophenes) and disulfides (e.g., methyl ethyl disulfide).

SulfaTrap LLC is offering a new sorbent-based desulfurization processes that can remove organic sulfur contaminants from gaseous and liquid hydrocarbon streams. The paper will provide the details of the new technology and the results of large scale pilot experiments showing the capabilities of processing natural gas liquids (C3/C4 streams) as well as feedstocks that are used in the chemical synthesis of value added products. A high fidelity economic assessment will provide the processing costs for the following cases:

- Case 1- Desulfurization of Off-spec C3/C4 Streams
- Case 2 â€“ Treatment of feedstocks

513113 Catalytic Reforming for Syngas Production Using Industrial CO_2: Carbon Capture and Chemical Manufacturing

Abstract:

Enerkem produces biofuels and renewable chemicals from waste. Its disruptive proprietary technology converts non-recyclable, non-compostable municipal solid waste into methanol and ethanol. Earlier this year, Enerkem expanded its Edmonton biofuel facility to produce some 13 million gallons of cellulosic ethanol annually following the commissioning of its methanol-to-ethanol conversion unit. In addition to ethanol, Enerkem plans to develop other widely-used chemical intermediates such as acetic acid, methyl acetate and acrylic acid through advanced catalysis. One of the recent developments is the scale-up of the Enerkemâ€™s reforming catalyst (E20) through Unicat Catalyst Technologies for generating CO-rich syngas for downstream carbonylation process. This syngas production route utilizes a dry reforming approach and converts waste CO_2 (procured either from the Enerkem process or any other fossil-based process) and natural gas to generate syngas with very stable performance. This increase in syngas production has the potential to improve the liquid products yield per bone-metric dry ton (BDMT) of the processed feedstock with significant carbon capture potential. The presentation will highlight the research on scaling the technologies developed to date from bench scale to pilot scale and on tailoring the process conditions of the technologies to facilitate integration into a bio-refinery platform.

ALKALI PROMOTED MOLYBDENUM SULFIDE CATALYSTS FOR THE PRODUCTION OF HIGHER ALCOHOLS BY FISCHER – TROPSCH SYNTHESIS

Belinda Molina[1], Mohammad Erfanul Raihan[2], Tracy J. Benson[2], and Russell Chianelli[1]*

[1]Department of Chemistry, Materials and Environmental Science and Engineering, University of Texas – El Paso, El Paso, TX

[2]Dan F. Smith Department of Chemical Engineering, Lamar University, Beaumont, TX

*Author to whom correspondence should be addressed

ABSTRACT

Fischer Tropsch Synthesis (FTS) is a well-recognized process for the production of hydrocarbons and higher alcohols from syngas. A novel hydrothermal synthesis technique has been explored for the production of alkali-promoted MoS_2 with and without cobalt addition for the production of medium chain (C6 – C8) alcohols. The catalysts were characterized using SEM, XRD, TEM, and BET to identify possible structure, active sites, and influence on activity and selectivity. Reaction experiments were performed in a fixed bed reactor to evaluate the catalytic activity for the formation of higher alcohols from synthesis gas in a range of experimental conditions that mirrors commercial production. Reaction results indicated that catalysts containing cesium alkali and cobalt promoters were the most stable, active, and selective for the formation of medium chain alcohols. The $Cs_{0.3}Co_{0.5}MoS_2$ catalyst (reacted at 3.2 MPa, 300 °C, and $H_2/CO = 0.8$) favored C6 – C8 alcohols (93 wt % total mass of products).

Keywords: Fischer – Tropsch Synthesis, Medium chain alcohols, Alkali moly-sulfide, Hydrothermal synthesis

1. Introduction

Fossil fuels have allowed technologies to advance in modern transportation for over a century. Over 80 million barrels of petroleum are processed every day in refineries worldwide to meet the need for liquid transportation fuels such as gasoline, diesel fuel and jet fuel. Despite the advancements made in increasing the fuel economy of vehicles, world usage of transportation fuels has continued to increase and is assumed to remain high.[1] Several realities emerged in the 20th century: the potential risks associated with the anthropogenic climate change, complex geopolitics, and energy security concerns such as the continuous availability of energy in varied forms, in sufficient quantities, and at reasonable prices and eventually the dwindling supply of this essential non-renewable resource.[2] For an analysis that extends well into the 21st century and explores the long-term availability of fossil fuels, the future production of unconventional oil and resources eases many of the concerns aforementioned.

Unconventional oil and resources include, but are not limited to, oil shale, heavy crude oil, tar sands (natural bitumen), coal, natural gas, and biomass. U.S. fuel improvement efforts have become focused on using alternative fuels to satisfy the sustainable development objectives of keeping fuels affordable, increasing energy security, and evolving towards near-zero emissions of both air pollutants and greenhouse gases.[3] Fischer Tropsch synthesis (FTS) provides a promising advantage of an efficient, synthetic route to ethanol and/or other higher alcohols. Over the last century there has been a continued search for the answers to this seemingly difficult task.

The original Fischer-Tropsch reaction is the heart of all gas-to-liquid technologies; it creates higher alcohols and hydrocarbons from CO/H_2 using a metal catalyst. The interest of this study is the selective formation of alcohols over MoS_2-based catalysts. More notably is the selection of compounds to accompany the, already well-recognized, catalytically efficient MoS_2. The MoS_2-based catalysts are attractive for various reasons. They withstand sulfur species in synthesis gas and allow less effort and cost to remove sulfur from syngas.[4 - 8] Moreover, alkali promoted cobalt molybdenum sulfides have previously shown to yield a substantial alcohol selectivity-even at a high CO conversion.[8 - 10]

To minimize or avoid the formation of undesired FT products and to allow the selective formation of alcohols, particularly medium to long chain alcohols (C5 – 16), it is necessary to use a catalyst that will promote the formation of surface intermediates that give rise to the production of alcohols. The catalyst should be able to change the thermodynamically controlled process, which favors the production of hydrocarbons, to a kinetically controlled process that favors the production of alcohols. Catalytic materials are strongly affected by every step of catalytic design and preparation. There has been substantial research regarding the catalytic effects of the compounds chosen, as optimal FT catalyst candidates. In this research; they are MoS_2, cesium and potassium as alkali promoters, and promotion by the Group VIII metal, cobalt.

Incorporation of Group VIII metals into the MoS_2 compound has significant effect on both the structure and the properties of the catalysts. Observed by Lauritsen et al. (2007), MoS_2 and Co-Mo-S have hexagonal structures.[11] In catalytic studies by Santiesteban (1989), alkali doping of MoS_2 resulted in higher yields of alcohols.[12] Early experiments indicated that alkali/MoS_2 materials provided for very specific product selectivity while remaining stable for long reaction runs.[13, 14] Of particular interest for this research, was identifying catalysts for efficient production of medium chain alcohols (C6 – C8).

It is important to mention that the novelty of these catalysts lies in the hydrothermal synthesis method, and that they are unsupported. The hydrothermal technique has offered several new advantages such as high reaction rate of powders, good dispersion in liquid, nearly pollution free, does not require expensive and highly sophisticated equipment, energy saving processing, and, in many cases, produces new phases. Detailed in this paper are the preparation methods, characterization, and catalytic testing of alkali-promoted molybdenum disulfides with the interest of exploring the fundamental explanations of the observed phenomenon.

2. Experimental

2.1. Catalyst Preparation

FT catalysts were synthesized using the hydrothermal method via a two-step precursor procedure as prescribed by Ornelas et al. (2002).[15] Synthesis began by preparing an ammonium tetrathiomolybdate (ATM) precursor.[15, 16] Ammonium heptamolybdate (12.5 g) was dissolved in 60 mL of deionized water before adding 100 mL of ammonium sulfide (20 wt % in H_2O). The solution was heated to 55 °C, and the temperature was maintained for 30 minutes under constant stirring. The products were a bright red ammonium salt that was then sealed and stored in a refrigerator until ready for use. The reaction is as follows:

$$20H_2O + (NH_4)_6Mo_7O_{24} + 4H_2O + 28(NH_4)_2S \rightarrow 7[(NH_4)_2MoS_4] + 48\ NH_4OH$$

Alkali, particularly K and Cs, promoted transition metal sulfide catalysts (MoS_2 and $CoMoS_2$) were synthesized, and the catalytic promotional effects were compared to that of MoS_2 and $CoMoS_2$ catalysts. Catalysts containing K (or Cs) were synthesized by mixing the alkali (either potassium carbonate or cesium acetate) with molybdenum (molar ratio alkali: Mo = 0.3) with ATM in aqueous solution. Synthesis for K/Cs-MoS_2 began by calculating the mass of the alkali compounds (with a ratio of 0.3 mole of alkali metal to 1 mole of molybdenum at 5.0 g of ATM). The ATM solution was filtered to collect 5.0 g. The 5 g of ATM was re-suspended in 50 mL of DI water. The alkali was dissolved separately in 10 mL of DI water and then mixed together while stirring for 15 min. For catalysts containing cobalt, 2.8 g of cobalt nitrate hexahydrate was dissolved in 10 mL of DI water and then added to the mixture last while stirring for an additional 15 minutes. For materials containing cobalt (cobalt nitrate hexahydrate), a Co:Mo = 0.5 was used, with the alkali:Mo remaining at 0.3. The catalyst materials were reacted in a 600 mL Parr reactor for 2 hrs at 573 K and 9 MPa. After the hydrothermal process, each catalyst was vacuum filtered, washed with isopropanol, and dried in a desiccator.

2.2. Characterization of Catalysts

X-ray Diffraction (XRD) scans were obtained using a Bruker D8 Discover and were conducted with a step size of 0.02 in 2θ at a scan rate of 5.0/min. Scanning electron microscopy (SEM) was performed using a Hitachi S- 4800 II SEM with EBSD/OIM. Specimens were ground in a mortar and pestle prior to mounting. A sample size of ~ 0.5 mg was mounted onto a 32 mm thin plate on an electrically-conductive double-sided carbon adhesive. Transmission Electron Microscopy (TEM) was performed using a Zeiss EM10 microscope. TEM samples were prepared by diluting a minimal amount of sample in isopropanol and sonicating for 30 minutes. A drop of the solution was placed on slot copper grid supported on formvar/carbon.

The surface area of the fresh and spent catalytic series was determined using the conventional multi-point BET method. The measurements were performed on a Micromeritics TriStar 3000 BET Surface and Porosity Analyzer. Prior to analysis, the samples were evacuated overnight and degassed with dry N_2 at 323 K for 1 hr.

2.3. Catalyst Testing

The FTS reactions were carried out in a packed bed reactor (5 mm ID, 316 SS) placed inside a tube furnace. The fixed bed reactor was operated in continuous, steady state mode and was flow controlled (~30 mL/min) and back-pressure regulated. Typical reaction pressures were 2.1 – 3.6 MPa. The tube was packed with 250 mg of catalyst with glass wool on either side of catalyst. Flow rates were adjusted for 3,000 to 7,500 GHSV (mL/(g_{cat}*hr)). Catalyst materials were reduced in situ prior to reaction by flowing H_2 at 200 °C for 2 hrs. Reactant concentrations 0.8 to 1.0 (H_2:CO) were used. Empty bed reactions were performed, resulting in no appreciable conversion of the syngas.

Reaction products were determined using a Varian 3800 GC equipped with FID. Product compounds were separated using an Rxi-1 ms capillary column (Restek, Inc.; Bellefonte, PA, USA). Analytical standards (C1 – C10 hydrocarbons, C1 – C8 alcohols (linear and branched), formic acid, acetic acid, and isobutyric acid) were used to identify and quantitate product compounds. All analytical standards were purchased from Sigma Aldrich and were used as received.

3. Results

3.1. Catalyst Characterization

The MoS_2 diffraction pattern exhibited the properties of a poorly crystalline hexagonal MoS_2. The initial peak to the left of the diffraction pattern (10 – 15 2θ) is the most predominant; this featured peak is consistent throughout the potassium series. Santiesteban et al. (1989) stated that this peak is broadly intense giving a strong indication that there is stacking of the MoS_2 layers.[12] The next intense peak occurring at 33 2θ provided evidence of these stacking faults as indicated by its asymmetric shape.

There were two phases present throughout these diffraction patterns, 1) disordered MoS_2 and 2) crystalline Co_9S_8. Chianelli, et al. (1979) mentioned that the broad intensity of the peak occurring at 38.7 2θ for the $K_{0.3}MoS_2$ and MoS_2 XRD patterns suggests that the two-layer MoS_2 stacking sequence is maintained for at least two stacks.[17] For the diffraction patterns of MoS_2 and $K_{0.3}MoS_2$ a broad MoS_2 peak is identified at 38.7 2θ with a d-spacing of 2.32 nm. When cobalt is present in the samples $Co_{0.5}MoS_2$ and $K_{0.3}Co_{0.5}$ MoS_2, a crystalline peak for CoS_2 occurs in proximity at 39.6 2θ with a d-spacing of 2.25 nm. The broad peak occurring between 55 – 60 2θ is the last indicated peak shown for the MoS_2 phase (Figure 1).

All of the materials show disordered MoS_2 with crystalline cobalt sulfide. For the potassium series catalyst, the addition of potassium to the precursor likely leads to segregated cobalt suggesting that thermally decomposed sulfide precursors may contain a mixture of molybdenum and cobalt sulfides instead of a dispersed Co-Mo-S type of material. The Co-Mo-S type phase was not present in these diffraction patterns. According to Chianelli, et al. (1985) the structure and performance of the subsequent finished catalyst is understood to be dependent on the intensity of the thermal treatment. Studies on phase equilibria under hydrothermal conditions have found that this method of catalytic synthesis may in fact give rise to new phases that do not have natural analogues or have not been identified.[18, 19] Between temperatures of 300 – 350 °C materials develop a greater abundance of crystalline cobalt sulfides, and on salted samples, potassium sulfides are present. An interesting feature of the XRD patterns was that no peak for potassium compounds were found. Possible reasons may be due to the high dispersion capability of the hydrothermal synthesis, preventing isolated peaks of the alkali promoter to form.[20 - 22]

Similar characteristics from the potassium series were observed in the XRD pattern of the cesium promoted series. The hexagonal phase of MoS_2 was still predominant showing the same corresponding peaks for MoS_2 phase. Peaks of CoS_2 were also detected at the same 2θ range.

Similar behavior was observed on $Co_{0.5}MoS_2$ and $Cs_{0.3}Co_{0.5}MoS_2$ between 20 – 27 2θ by Santiestaban (1989). He attributed this to exposure to the reaction atmosphere during catalytic tests for long periods of time showing the agglomeration and crystallization of the cesium compound (Cs_2CO_3).[12] This phase was not recognized by the analytical methods by which peak identification was performed. An interesting feature recognized in this diffraction pattern is the shifting to the left when $Co_{0.5}MoS_2$ is promoted with cesium. The shift to a lower diffraction angle is due to the insertion of cesium between CoS_2 and MoS_2 sheets and the expansion of the interlayer spacing (d-spacing), the shift of the peaks after adding the alkali was also described by Ikeda and coworkers.[21]

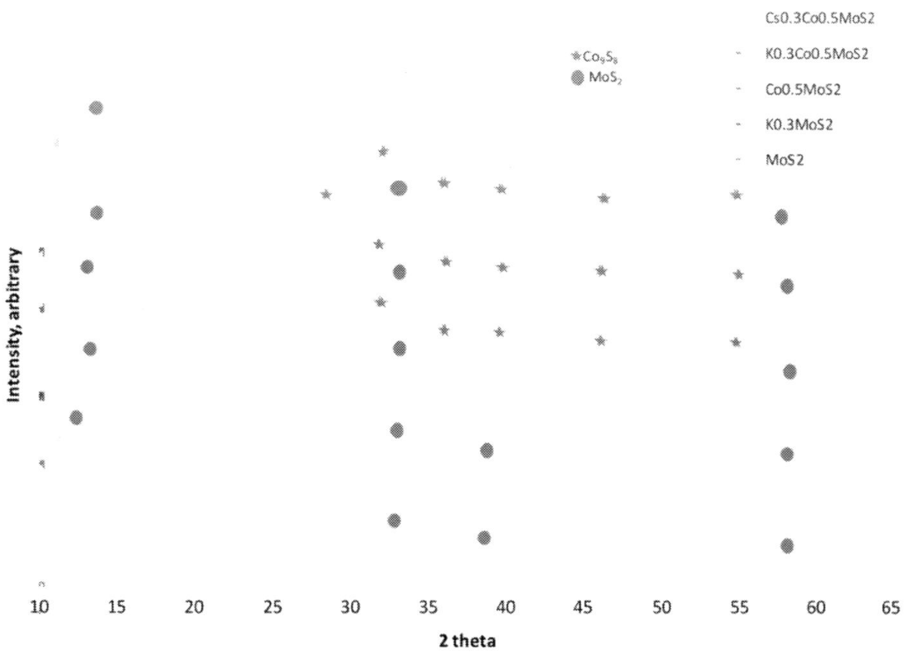

Figure 1. X-Ray diffraction pattern and peak identification for the MoS_2 Promoted Catalysts.

X-ray diffraction patterns of spent catalysts (Figure 2) show what appears to be carbide formations. For the $KCoMoS_2$ catalyst, the appearance of newly formed peaks, compared to the fresh catalyst, at 43.2 and 50.2 2θ can be attributed to the formation of Co-Mo-C and K-Mo-C, respectively. Xiang et al. (2007) identified K-Mo-C as an active site for alcohol formation. For the $CsCoMoS_2$ catalyst, peaks at 24.5 and 30.5 2θ give rise to the possibility for agglomeration of cesium on the catalytic surface.[23 - 26] In addition, the disappearance of the Co-S peaks (32.2, 36.2, 39.8, 46.3, 54.9 2θ) in the spent catalyst could support the idea for a change in the state of the cobalt promoter, perhaps being incorporated into the Mo-S phase; however, this is purely speculative.

As can be seen in Table 1, the addition of alkali promoters to MoS_2 led to a surface area increase. When potassium was added to MoS_2, it increased the surface area by ~ 100 m^2/g (or a factor of 1.6). However, when cesium was added, the surface area increased by a factor of 2.63. The alkali metal insertion into the MoS_2 expands the distance between the MoS_2 stacking, giving a higher surface area. Catalytic surface area varies with the ionic radius of the alkali metals.[27] It is possible that the XRD analysis did not observe peaks of alkali metal phases due to the high dispersion of this alkali metal. The surface area for $Co_{0.5}MoS_2$ was 66 m^2/g. Upon addition of potassium, the surface area remained nearly the same. Alkali metals occupy the surface porosity. Potassium's ionic radius is smaller than that of cesium; therefore, the potassium enters the pore but does not completely block it. Hence, the surface area remained nearly the same. Cesium on the other hand, having a larger ionic radius, leads to a loss of surface area.

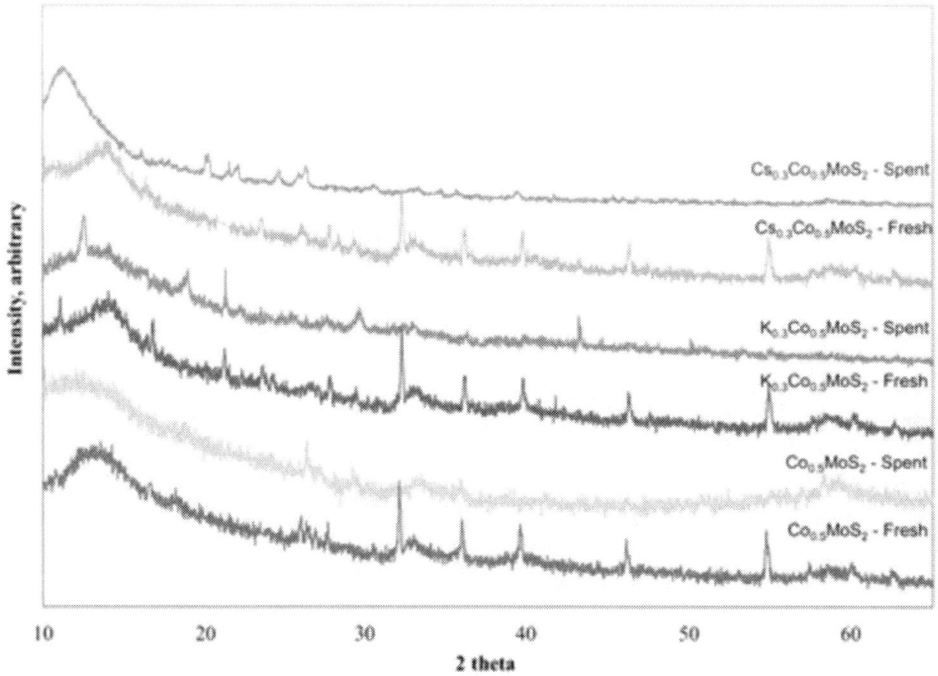

Figure 2. X-Ray Diffraction patterns of fresh vs. spent catalysts.

BET surface area measurements were also taken for the fresh versus spent catalyst (Table 1). The majority of the catalysts had a surface area decrease when comparing fresh versus spent, which is common for catalysts exposed to syngas under high pressure and temperature. The loss in surface area of the studied catalysts was quite significant. One possible reason is that the alkali promoter could have agglomerated onto the surface during the reaction. This agglomeration, in turn, could have blocked the pores of the catalyst, thus giving a lower surface area.[28, 29] The decrease in surface area was the case for all the catalysts, except $K_{0.3}Co_{0.5}MoS_2$. This catalyst observed a dramatic increase in surface area after exposure to syngas. To help explain this phenomenon, large pore distribution enables gaseous reactant molecules, during adsorption, to enter and exit out of the pore site. This allows products to be adsorbed and dispersed onto the catalytic surface as well as desorbed (i.e. the "build up" of products do not block pore and surface sites). Product build up on pore sites is not a favorable characteristic of an effective catalyst. In addition, it has been previously shown that when Transition Metal Sulfide catalysts "line out" (i.e. activity is stabilized), the catalyst particles stabilize with a novel carbide surface layer. The amorphous materials are similar to glasses with low surface areas.[17, 19]

Table 1. Surface area and compositional analysis for fresh and spent catalyst

Catalyst Sample	Area, m^2/g (Fresh)	Area, m^2/g (Spent)
MoS_2	169	
$K_{0.3}MoS_2$	268	
$Cs_{0.3}MoS_2$	444	41
$Co_{0.5}MoS_2$	66	23
$K_{0.3}Co_{0.5}MoS_2$	70	481
$Cs_{0.3}Co_{0.5}MoS_2$	33	16

Two features were observed using the SEM and TEM. A feature that was observed consistently in the alkali-$Co_{0.5}MoS_2$ series and alkali-MoS_2 series was MoS_2 stacking. The SEM image shown in Figure 3 was taken at an angle of a particle observing MoS_2 stacking on the surface as well as on its side angle. This has been observed as the "rag" structure consisting of several stacked and highly disordered S-Mo-S layers. XRD analysis confirms poorly crystalline structure of hexagonal MoS_2.[30]

Figure 3. MoS_2 Stacking observed through SEM.

TEM images show how hexagonal slabs of MoS_2 agglomerate together to form arbitrary shapes (Figure 4, a), a feature that was also observed by Baksh (2004).[30] The sulfide has a hexagonal structure with Mo atoms placed in the trigonal-prismatic holes in the plane between the sheets of sulfur. If synthesized in a more hydrogen rich environment, the adsorption of hydrogen at the edges serves to stabilize the S edge relative to the Mo edge, and for this reason, the sulfide particle attains a more truncated hexagonal shape that exposes both edge types. Taster and Chianelli (1980) suggested that the MoS_2 (0001) basal plane is essentially inert and that test molecules, such as NO, CO, and O_2, have revealed that the catalytic activity of MoS_2 is related to the edges of the sulfide.[32 - 34] The second feature observed by TEM was a MoS_2 "nano-flower" morphology (Figure 4, b). The X-ray diffraction patterns of the spent catalyst samples could support the notion of a change is the state of the cobalt promoter ($Co_{.5}MoS_2$) since the appearance of an agglomerated morphology (see SEM images in Figure 5) is associated with a decrease in surface area (66 to 23 m^2/g).

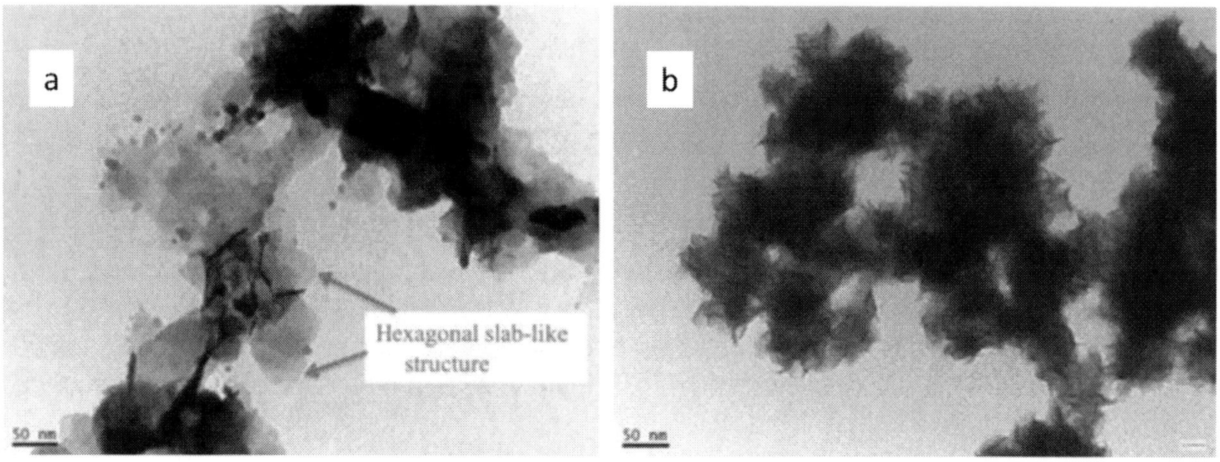

Figure 4. TEM micrographs illustrating a) Slab like MoS_2 hexagonal structures for the $K_{0.3}Co_{0.5}MoS_2$ catalyst and b) Nano flower structure for MoS_2.

Figure 5. SEM images of the $Co_{0.5}MoS_2$ before and after FT synthesis reactions.

3.2. Catalytic Activity and Selectivity

The ability for a catalyst to participate in the Fischer Tropsch Synthesis and yield the desired activity and selectivity depends on many factors. Among these factors are temperature, H_2/CO ratio, pressure, space velocity, and catalyst composition. Initial experiments were conducted using the MoS_2 catalyst to establish appropriate reaction conditions (i.e. 3.2 MPa, 300 °C, H_2/CO = 0.8, and GHSV = 5,000). A representative composition profile for the catalysts studied can be found in Figure 6. An interesting feature found in all the catalysts is a "lining up" effect. Low molecular weight hydrocarbons are predominantly produced at the beginning of the reaction run (t = 0 hrs). By the end of the first hour of time on stream (t = 1 hr), the low MW HCs become minor products. By the 5th hour, product selectivities and conversions appear to mostly stabilize. The results shown in Figure 6 are only meant to illustrate how a change occurs within the catalyst during the initial hours of on stream reaction results in a change in products, as evidenced by Christensen, et al. (2009). The experiments were not conducted (at least for these results) to offer any evidence towards a reaction scheme. It should be noted that in this work CO conversions were intentionally kept low with the intent to piece together a plausible reaction scenario, as illustrated by Benson and co-workers (2007 and 2009). [35,36]

As can be seen from Table 2, product selectivities were indeed related to catalyst composition. Note that in this comparison of catalysts, all reactions were performed at identical conditions (i.e. 3.6 MPa, 300 °C, and H_2/CO = 0.8). Catalysts containing cobalt yielded alcohol products in the C5 – C8 range, and only the normal chain compounds were observed. Low molecular weight hydrocarbons appeared in small concentrations for catalysts containing cesium. Catalysts containing potassium, without the aid of cobalt, produced n-butanol as well as branched chain pentanol isomers. To help explain these results, one must look at the reactivity of the surface species. Several surface science studies have investigated the deposition of alkali metals on MoS_2. A low coverage of an alkali metal on MoS_2 results in a disordered overlayer of strongly ionized adspecies. At higher coverage, the alkali adspecies begin to form 2D islands and additional alkali deposition yields 3D structures.[34, 37, 38] For alkali adsorption on the WS_2 (0001) surface, for example, which has the same structure as MoS_2 (0001), it was observed that the alkali addition causes the largest attenuation of the tungsten signal in ion scattering spectroscopy.[32] This has been interpreted in the way that the alkali adatoms occupy the trifold hollow sites directly above the metal atoms.[39 - 40] DFT calculations suggest that this is also the preferred adsorption site on MoS_2.[39, 40] Without an alkali promoter, the sulfide produces essentially no alcohols in CO hydrogenation, regardless of the conversion level.[41] Furthermore, the addition of an alkali promoter to the sulfide causes a decrease in hydrotreating activity and a very significant decrease in the activity for both hydrogenation of ethylene and methanation of CO.[39] Very importantly, as shown by many researchers, the addition of the alkali promoter shifts the product distribution in high pressure CO hydrogenation from hydrocarbons towards alcohols.[23 - 26]

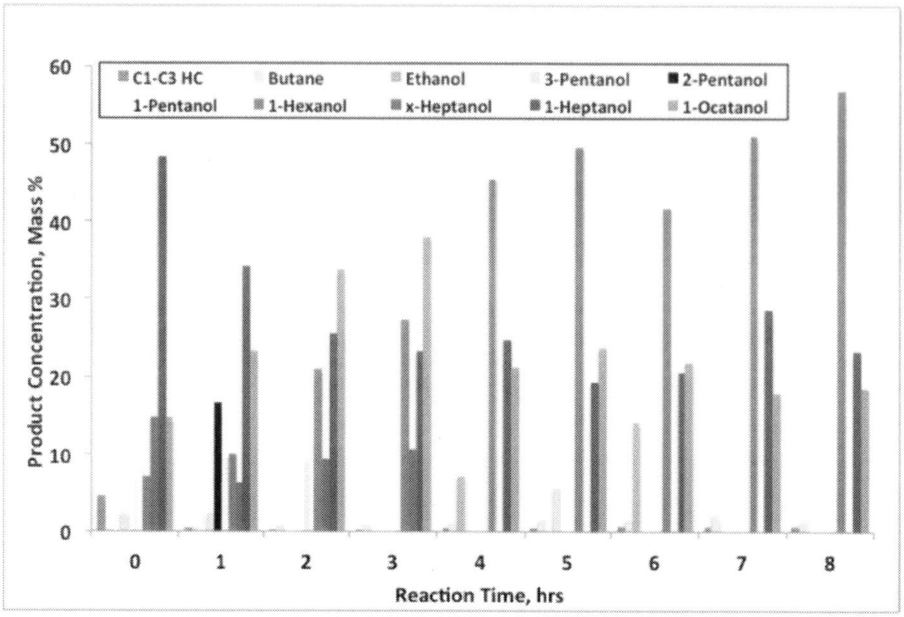

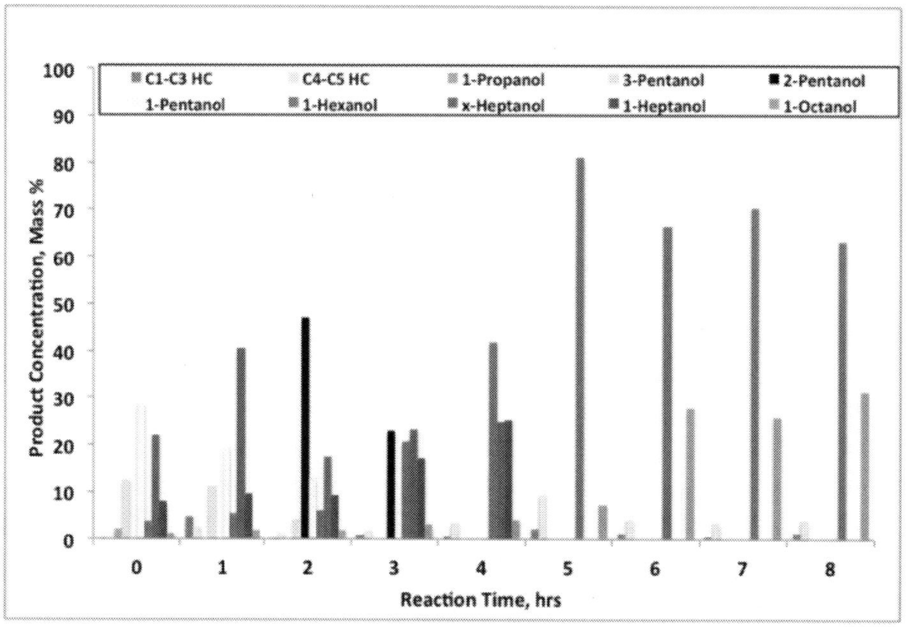

Figure 6. Product distribution of catalyst $Cs_{0.3}Co_{0.5}MoS_2$ (top) and $Co_{0.3}MoS_2$ (bottom) illustating the change in products from the initial four hours throughout the remainder of the reaction run (3.2 MPa, 300 °C, $H_2/CO = 0.8$, and GHSV 5,000 mL/(gcat*hr).

Table 2. Average product composition (µg/mL) from GC analysis for FT reactions occurring at 3.6 MPa, 300 °C, and $H_2/CO = 0.8$ (values in parentheses designate one standard deviation)

	MoS_2	$CoMoS_2$	$KMoS_2$	$CsMoS_2$	$KCoMoS_2$	$CsCoMoS_2$
C1 –C3 HC				0.1 (0.05)		0.2 (0.01)
C4 – C5 HC				0.53 (0.18)		0.42 (0.12)
2-Propanol	4.8 (0.44)					
1-Butanol			5.6 (0.77)			
2-Pentanol			7.0 (4.33)			
3-Pentanol			10.1 (2.42)			
1-Pentanol		11.3 (0.82)				
1-Hexanol	9.8 (2.34)	14.0 (2.09)		7.35 (0.558)	8.08 (2.36)	13.59 (3.04)
1-Heptanol		7.9 (0.375)			8.24 (2.77)	25.85 (1.08)
1-Octanol				2.58 (1.40)	5.42 (1.33)	5.26 (1.83)

Product compositions of various catalysts were compared based on different reaction conditions. Early experiments concluded that pressure and H_2/CO played important roles in product distributions for the various catalysts. In an effort to develop a reasonable pathway for the products formed, reaction pressures and H_2/CO ratios were adjusted to elucidate the initial products formed during the reaction sequence. FT synthesis reactions are complicated and are comprised of series of short-lived intermediate reaction products. In a typical FT synthesis reaction, both hydrocarbons and alcohols are produced. However, one of them is a major product, and the other remains as side product. Initially, in all FT reactions, intermediates are produced and further depending upon the catalyst and reaction conditions, these intermediates go to chain propagation or early chain termination to produce chiefly hydrocarbons and alcohols. To optimize alcohol synthesis in an FT reaction, such catalysts and operating conditions are desired that can offer maximum intermediates to produce alcohol compounds rather than hydrocarbons. To shift the equilibrium towards alcohol, rather than hydrocarbon, products, the H_2/CO ratio was kept at 0.8. When a 1:1 ratio was used, product formation followed that typically found in the literature using similar types of catalysts.[23 - 26]

As the FT reaction proceeds, water is produced via the water gas shift (WGS) reaction. The short chain alcohol intermediates may go to early termination reaction by reacting with water to produce organic acids (i.e. formic acid and acetic acid) only when the concentrations of H_2O in the reaction mixture is sufficient to oxidize these intermediates. MoS_2 catalysts provide higher WGS activities, which reduce the H_2O concentration, and H_2 and CO_2 are produced at the expense of CO.[42 - 47] When FT synthesis produces considerable amounts of organic acid compounds (i.e. formic acid and acetic acid), it was hypothesized that lower WGS activities occur and resulted in early termination of alcohol intermediates. The loss of alcohol intermediates to organic acid is considered an obstacle towards the propagation to longer alcohol chains. For example, $Co_{0.5}MoS_2$ catalyst (3.6 MPa, H_2/CO = 1.0, and 300 °C) produced ~ 40 wt % of butane, pentane, and acetic acid (Figure 7). Note the change in H_2/CO ratio as compared to the results presented in Table 2 for the $CoMoS_2$ catalyst. The equilibrium product may be adjusted via the H_2/CO ratio. The loss of intermediates due to the generation of these compounds is not considered as an ideal condition to optimize alcohol production. Whereas, $Cs_{0.3}Co_{0.5}MoS_2$ catalyst (3.2 MPa, H_2/CO = 0.8, and 300 °C) does not produce organic acids, implying higher WGS reaction activities, and produces hydrocarbons (~ 7 wt %) as side product (i.e. favors alcohol selectivity). Medium chain alcohols (C6 – C8) were found ~ 93 wt % of the total mass of products. Therefore, the $Cs_{0.3}Co_{0.5}MoS_2$ catalyst under the specified condition is feasible for optimum medium chain alcohol production.

Dry (1990) proposed a mechanism for the overall pathway for hydrocarbons and alcohols production from syngas based on a MoS_2 based catalyst.[43] In his mechanism, he had shown both ~CH_2 and CO as the active species, where for alcohols production, the active species ~CH_2 takes part in chain propagation and CO produced the alcohol intermediates. From the experimental analysis shown here, most of the products were normal chain alcohols with a few hydrocarbons as well as, in some cases, organic acids. In addition, these reactions consisted of a number of short-lived intermediates. As these modified FT catalysts were shown to undergo complex series of reactions, it was not possible to draw the exact mechanism. However, based on the alcohols, hydrocarbons, and organic acids produced, an overall reaction mechanism is proposed (Scheme 1).

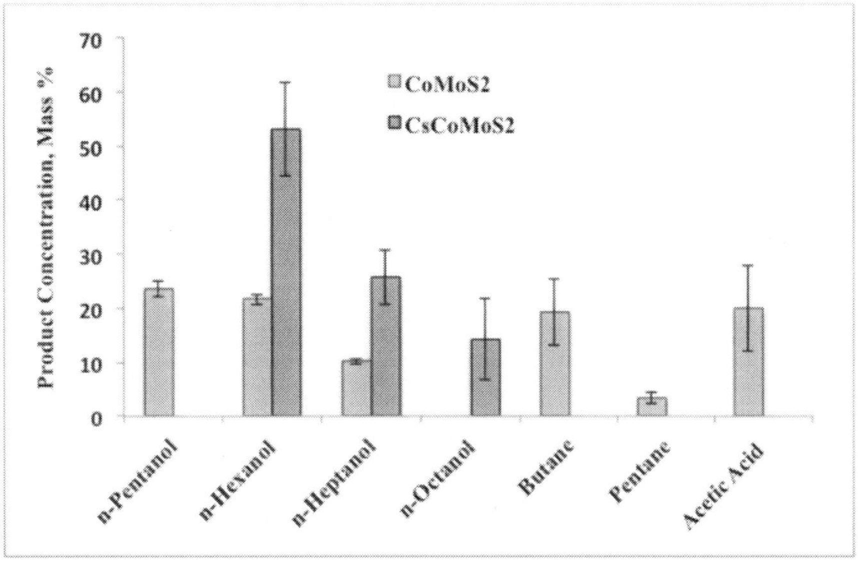

Figure 7. Effects of Water Gas Shift on Alcohol Selectivity. ($Co_{0.5}MoS_2$ operated at 3.6 MPa, $H_2/CO = 1.0$ and $Cs_{0.3}Co_{0.5}MoS_2$ operated at 3.2 MPa, $H_2/CO = 0.8$).

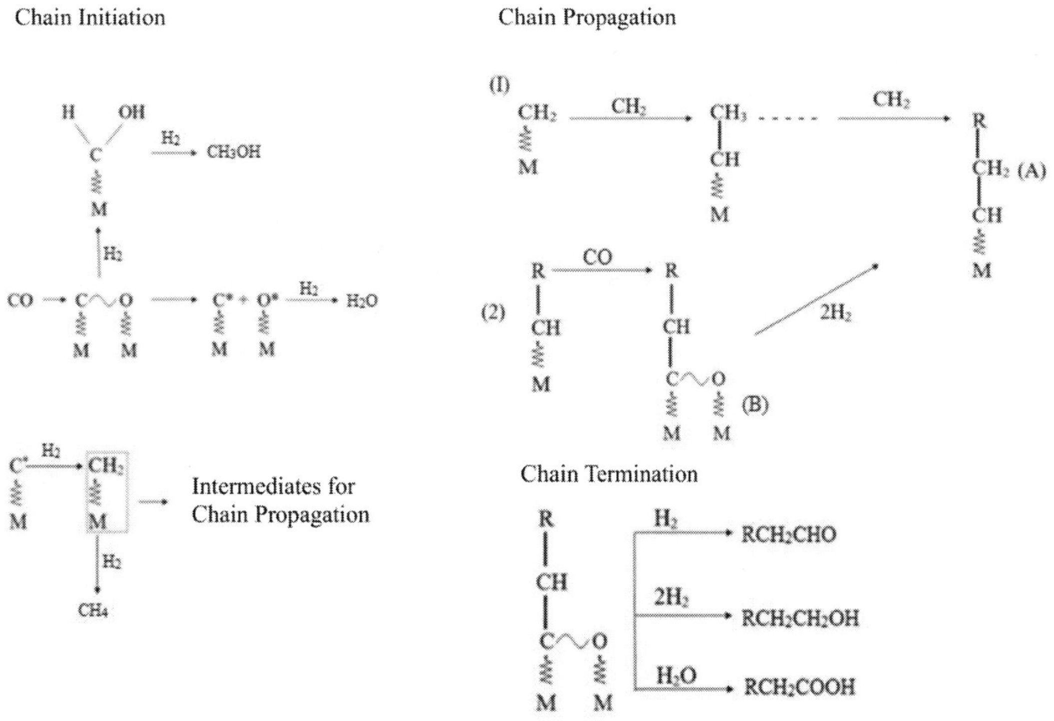

Scheme 1. Proposed overall reaction pathway for FT Synthesis of medium chain alcohols.

4. Conclusions

This research contributes to the understanding of hydrothermally synthesized unsupported alkali-promoted MoS_2- based catalysts. The XRD patterns for the potassium and cesium series observed poorly crystalline phases of MoS_2. These peaks, in conjunction with TEM and SEM analysis, were attributed to MoS_2 stacking. CoS_2 peaks were also consistently observed in both the cesium and potassium phases, owing to a segregation of the cobalt during thermal decomposition of the sulfide precursors instead of Co-Mo-S materials. The shift in the diffraction angle for potassium and cesium series catalysts could be due to the insertion of the alkali between CoS_2 and MoS_2 sheets and the expansion of the interlayer spacing (d-spacing) or could be due to structural changes of intercalation, which is not completely understood and is a subject of interest for future work. Fresh versus Spent x-ray diffraction patterns concluded that MoS_2 peaks dissipated over the syngas exposed catalysts. It is possible that agglomeration of cesium occurred on the catalytic surface. Phase equilibria studies on hydrothermal conditions have found that this method of catalytic synthesis may in fact give rise to new phases that do not have natural analogues or have not been identified in prior literature.

BET porosity and surface analysis concluded that alkali addition to MoS_2 increased surface area. Spent versus fresh surface areas are consistent with that reported in literature. There was a decrease in surface area for most syngas exposed catalysts, perhaps due to agglomeration of catalytic site species at reaction conditions. However, the $K_{0.3}Co_{0.5}MoS_2$ catalyst showed an increase in surface area once it was exposed to syngas (70 m^2/g compared 481 m^2/g). SEM and TEM electron microscopy confirmed two morphological features present on the catalytic surface: the hexagonal structure consisting of several stacked and highly disordered S-Mo-S layers and the MoS_2-nanoflower.

Optimum FT reaction conditions for medium chain alcohol products were 3.6 MPa, 300 °C, and $H_2/CO = 0.8$ using alkali-promoted $CoMoS_2$ catalysts. The $Cs_{0.3}Co_{0.5}MoS_2$ was the most stable, active, and selective towards medium chain alcohol formation. The alkali-promoted $CoMoS_2$ catalysts produced the least diversity of alcohol groups, indicating a high degree of selectivity. In addition, this work demonstrates the importance of the syngas composition (i.e. H_2/CO) on product composition. $H_2/CO < 1$ yielded predominantly alcohol products, rather than hydrocarbons; however, one can speculate that doing so would cause a reduction in CO conversion.

Acknowledgements

The authors would like to thank Drs. Brenda Torres, Maryam Zarei, and Mahsa Hosseini from the University of Texas at El Paso's Materials Research and Technology Institute. This research was funded in part by NASA (Grant #: NNX10AL46H). Transmission Electron Microscopy images were sponsored by grant BBRC NIMHD5G12MD007592. The authors would like to thank the Lamar University College of Engineering for assistance in reactor equipment and GC analysis.

References

[1] Kaufmann, T.G. Kaldor, A., Stuntz, G.F., Ansell, L. L. (2000) *Catal. Today*, 62, 77.
[2] de La Ree, A. B., Best, L. M., Bradford, R. L., Gonzalez-Arroyo, R., Hepp, A. F. (2012) *The NASA Scientific and Technical Information Program*, Serial number 217402.
[3] "World Energy Assessment: Energy and the Challenge of Sustainability." (2000) United Nations Development Program. http://www.undp.org/content/undp/en/home/librarypage/environment-energy/sustainable_energy/world_energy_assessmentenergyandthechallengeofsustainability.html (accessed September 2017)
[4] Liu, Z., Li, X.,Close, M. R., Kugler, E. L., Petersen, J. L., Dadyburjor, D. B. (1997) *Ind. Eng. Chem. Res.* 36, 3085.
[5] Murchison, C. B., Conway, M. M., Stevens, R. R., Quarderer, G. J. in: Phillips, M. J., Ternan, M. (Eds.) (1988) *Proceedings of the 9th International Congress of Catalysis*, Vol. 2, Calgary, Chemical Institute of Canada, Ottawa, p. 626.
[6] Bian, G., Fan, L., Fu, Y., Fujimoto, K. (1998) *Ind. Eng. Chem. Res.*, 37, 1736.
[7] Gang, L., Chengfang, Z., Yanqing, C., Zhibin, Z., Yianhui, N., Linjun, C., Fong, Y. (1997) *Appl. Catal. A*, 150, 243.
[8] Dianis, W. P. (1987) *Appl. Catal.*, 30, 99.
[9] Stevens, R. R., Conway, M. M. (1989) US Patent 4,831,060.
[10] Qi, H., Li, D., Yang, C., Ma, Y., Li, W., Sun, Y., Zhong, B. (2003) *Catal. Commun.*, 4, 339.
[11] Lauritsen, J.V., Kibsgaard, J., Olesen, G. H., Moses, P.G., Hinnemann, B., Helveg, S., Nørskov, J. K., Clausen, B. S., Topsøe, H., Lægsgaard, E., Besenbacher, F. (2007) *J. Catal.*, 249, 220.
[12] Santiesteban, J. G. (1989) *Alcohol Synthesis from Carbon Monoxide and Hydrogen over MoS_2-Based Catalysts*. Thesis. Lehigh University.
[13] Nava, H., Ornelas, C., Aguilar, A., Berhault, G., Fuentes, S., Alonso, G. (2003) *Catal. Let.*, 86, 257.
[14] Surisetty, V. R., Dalai, A.K. Kozinski, J. (2011) Appl. Catal. A, 404, 1.
[15] Nava, H., Ornelas, C., Aguilar, A., Berhault, G., Fuentes, S., Alonso, G. (2002) *Catal. Lett.*, 86, 257
[16] Andersonn, R., Boutonnet, M., Jaras, S. (2014) *Fuel*, 115, 544.
[17] Chianelli, R. R., Prestridge, E. B., Pecoraro, T. A., DeNeufville, J. P. (1979) *Science* (Washington, DC, U.S.) 203, 1107.
[18] Byrappa, K., Yoshimura, M. (2001) Ch. 1: Hydrothermal Technology – Principles and Applications. Handbook of Hydrothermal Technology: A Technology for Crystal Growth and Materials Processing. Norwich, NY: Noyes Publication.
[19] Daage, M., Chianelli, M. (1994) *J. Catal.* 149, 414.
[20] Menart, M. J., Hensley, J. E., Costelow, K. E. (2012) *Appl. Catal., A*, 36, 437.
[21] Ikeda, S., Hara, M., Kondo, J. N., Domen, K., Takahashi, H., Okubo, T., Kakihana, M. (1998) *J. Mater. Res.*, 13, 852.
[22] Iranmahboob, J., Toghiani, H., Hill, D. O. (2003) *J. Appl. Catal. A*, 247, 207.
[23] Xiang, M., Li, D., Li, W., Zhong, B., Sun, Y. (2007) *Catal. Comm.*, 8, 503.
[24] Xiang, M., Li, D., Qi, H., Li, W., Zhong, B., Sun, Y. (2007) Fuel, 86, 1298.
[25] Christensen, J. M., Mortensen, P. M., Trane, R., Jensen, P. A., Jensen, A. D. (2009) *Applied Catal. A: Gen.*, 366, 29.

[26] Wang, J. J., Xie, J. R., Huang, Y. H., Chen, B. H., Lin, G. D., (2013) *Appl. Catal. A: Gen.*, 468, 44.
[27] He, J., Zhang, W., Zhejiang, J. (2008) *University Science*, 9, 714.
[28] Storch, H. H., Columbic, N., Anderson, R. R. (1951) *RSC Catal.*, 5, 48.
[29] Curtis, C. W., Pellegrino, J. L. (1989) *Energy Fuels*, 3, 160.
[30] Baksh, F. (2004) Synthesis Gas Conversion to Aliphatic Alcohols: Study of MoS_2 Catalytic Systems. Thesis. University of Punjab.
[31] Taster, S. J., Pecoraro, T. A., Chianelli, R. R. (1980) *J. Catal.*, 63, 515.
[32] Qingfeng, Z., Yu, K., Zhao, B., Wang, Y., Song, C., Li, A., Yin, H., Zhang, Z., Zhu, Z. (2013) *RSC Adv.*, 3, 10994.
[33] Kantschewa, M., Delannay, F., Jeziorowski, H., Delgado, E., Eder, S., Ertl, G., Knözinger, H. (1984) *J. Catal.*, 87, 482.
[34] Papageorgopoulos, C., Kamaratos, M., Kennou, S., Vlachos, D. (1991) *Surf. Sci.*, 251, 1057.
[35] Benson, T. J., Hernandez, R., French, T., Alley, E., Holmes, W. (2007) *J. Molec. Catal. A: Chem.*, 274, 173.
[36] Benson, T. J., Hernandez, R., French, W. T., Alley, E. G., Holmes, W. E., (2009) *J. Molec. Catal. A: Chem.*, 303, 117.
[37] Kennou, S., Ladas, S., Papageorgopoulos, C. (1985) *Surf. Sci.*, 152, 1213.
[38] Ohuchi, F. S., Jaegermann, W., Pettenkofer, C., Parkinson, B. A., (1989) *Langmuir*, 5, 439.
[39] Kantschewa, M., Delannay, F., Jeziorowski, H., Delgado, E., Eder, S., Ertl, G., Knözinger, H. (1984) *J. Catal.*, 87, 482.
[40] Youchang, X., Naasz, B. M., Somorjai, G. A., (1986) *Appl. Catal.*, 27, 233.
[41] Andersson, R. (2015) Catalytic Conversion of Syngas to Higher Alcohols over MoS_2-Based Catalysts, Thesis, KTH Royal Institute of Technology, Stockholm, Sweden.
[42] Newsome, D. S. (1980) *Catal. Rev.: Sci. Eng.*, 21, 275.
[43] Dry, M. E. (1990) *Catal. Today*, 6, 183.
[44] Fang, K., Li, D., Lin, M., Xiang, M., Wei, W., Sun, Y. (2009) *Catal. Today*, 147, 133.
[45] Surisetty, V. R., Dalai, A. K., Kozinski, J. (2010) *Ind. Eng. Chem. Res.*, 49, 6956.
[46] Surisetty, V. R., Dalai, A. K., Kozinski, J. (2011) *Energy Fuels*, 25, 580.
[47] Hindermann, J. P., Hutchings, G. J., Kiennemann, A. (1993) *Catal. Rev. – Sci. Eng.*, 35, 1.

511868 How to Sell the Value of Digitalization with MPC

Abstract:

Often engineers struggle with selling the value of their ideas on solving problems. This can be compounded when the solution involves new digital technology, like embedded model predictive control (MPC). This presentation will focus on a real world example using new MPC technology and how to justify and sell the solution to management.

513258 *Autonomous Tuning and Firing Temperature Control for Variable Engine Operating Conditions - a New Control Scheme Designed for Optimized Engine Control in All Seasons and a Variety of Fuel Gas Compositions*

Abstract:

Abstract

Siemens has developed a gas turbine control optimization platform called Gas Turbine Autonomous Control Optimizer (GT-ACO), designed to optimize various gas turbine control functions and deliver enhanced value to our customers.

With the advent of premixed dry low NOx and ultra-low NOx combustion systems NOx, CO and particulate emissions from gas fired power generation turbines have been reduced. However, premixed combustion systems have presented challenges, one of which is addressing combustion dynamics. Combustion dynamics are driven by rapid pressure fluctuations within the combustor. A pilot or diffusion flame is typically used to control combustion dynamics. However, to reduce emissions, the diffusion flame must be minimized. The first module of the GT-ACO 2.0 platform, Combustion Optimization, includes use of data-driven reinforcement learning technology to perform closed loop dynamics/NOx control optimization in order to moderate emissions while keeping combustion dynamics within targeted limits with no other changes to combustion hardware.

In addition, gas turbine controls have used exhaust temperature measurements correlated to turbine inlet temperatures to control the unit. Such methodologies typically rely on assumptions for cooling air flows, inlet and exhaust conditions, and can be influenced by transient conditions. Improvements can be made using a physics-based real time thermal model (RTTM) to calculate and control directly to turbine inlet temperature (TIT), which can reduce under- and over-firing conditions during engine operation and provide flexibility for operation with varying fuel gas content.

GT-ACO 2.0 has been developed as a platform to improve control of the gas turbine for the cases described above. However, there are many more modules that can be developed since it uses the Siemens Synalytics platform to transform big data into useable data in support of optimization. By applying such technologies, Siemens is can deliver more intelligent, proactive and customized services that can translate into real customer value through improved service offerings, customer experience, and introduction of the next generation of products.

This paper will dive deeper into the GT-ACO platform and describe how application of this new technology can help optimize gas turbine operation.

513335 To Cloud or Not to Cloud: That Is the Question

Abstract:

As companies seek value from their process data, selecting the appropriate infrastructure is of critical importance. At a minimum, one should consider security, features, scale and lifecycle cost for the four primary components of IIoT: data acquisition, analysis, historian and visualization. Here we discuss some of the trade-offs between on-premises and cloud components, with the aim of providing useful infrastructure and application decision criteria.

513813 The Journey Towards 2030, Was It to Ambitious? Update on the Impact of 2030 Vision and 2020 Transformation Program on Energy, Infrastructure and Other Projects in the Region

Abstract:

Saudi Arabia and other Gulf countries announced their vision for 2030 trying to have better future for next generations and prepare for the after oil ERA.

However, we have seen some countries as well as the World Bank calling for easing the transformation to avoid falling in a recession. In this paper we will see the effect of this vision and transformation program and other programs on the projects in the region during the past period and see how this effect the oil would and gas business and projects in the region? Will this mean more projects for the energy sector and petrochemicals or are countries moving away from there? Will low oil prices and budget deficits postpone or otherwise accelerate this transformation? If yes, then for which projects? Will we see more fuel expansion projects? Renewables? Solar? Will entertainment, infrastructure and housing projects have the highest priority?

512406 Refinery-Petrochemical Integration: Organization and Culture

Abstract:

Petrochemical plants and refineries both process hydrocarbon derived materials. Refineries focus on fuels production. Refinery products include LPG (liquefied petroleum gas), gasoline (petrol), kerosene and jet fuel, diesel, fuel oils, and coke. Petrochemicals focus on creating intermediates used to produce industrial and consumer products. Major petrochemical processes include steam cracking to produce ethylene (ethane), propylene (propene), and butadiene and aromatics plants producing benzene, toluene, xylene and other aromatics.

At intermediate points in their processes, refineries produce molecules of interest to petrochemical operations, and petrochemical plants produce molecules important to refineries. The easiest molecule to make is the one you already have. Refinery-petrochemical integration works on maximizing the return on the entire hydrocarbon value chain.

In refinery-petrochemical integration three different organizations are often present;

- refinery
- petrochemical plant
- transportation link or market maker between them

This presentation focuses on cultural and organizational differences between different entities and how that can make or break projects.

513278 One Stone Two Birds: CO_2 to Methanol and Petrochemicals

Abstract:

One stone two birds: CO_2 to Methanol and Petrochemicals.

Oil prices are making gasoline and naphtha production less attractive, and it is expected a net reduction in the long-term demand since new environmental friendly energy sources are becoming available and seem to be most attractive for the future generations. On another hand, the boom of Shale Gas set the path for the construction of new ethane crackers or the switch of flexible crackers to ethane, hence the available capacity of two main sub products from naphtha crackers and refineries will be substantially reduced, propylene and aromatics.

Technologies are available or evolving to fill this gap (i.e. Propane Dehydrogenation, Olefins Metathesis, Acetic Acid Hydrogenation) however these are not enough attractive by itself, an additional driver is needed to make new technology profitable and interesting; such is the case of Methanol. In recent years this product has been in the spotlight as a viable option to produce olefins and aromatics, i.e. in China where fuel resources come mainly from coal instead of oil and gas.

Improvement of catalysis to increase the selectivity to olefins or aromatics is making a great progress, but these processes are very demanding in terms of energy (i.e. better selectivity at high temperature or energy required for further separation.), this energy demand can be indirectly translated in a higher amount of CO_2 emissions which may become an obstacle for a further development of this alternative.

However, the chemical simplicity of Methanol makes it a flexible component, which can be produced from either fossil fuels or synthetic sources such as CO_2. Catalytic conversion of CO2 into Methanol using hydrogen is getting more attention and if this is feasible, the resulting synergy could be a simple solution for two different problems, the contribution for a sustained supply of aromatics and olefins with a positive environmental impact.

This paper presents pros and cons of this integration in terms of economic feasibility, environmental benefit.

512394 A New Era for Alkylation

Abstract:

In September 2016, Honeywell UOP introduced the refining industryâ€™s first new liquid alkylation technology in 75 years. ISOALKY^{TM} technology offers refiners the ability to produce alkylate using a specialized ionic liquid catalyst with lower handling risks and comparable process and economic performance to traditional hydrofluoric or sulfuric acid technologies.

Researchers have assumed ionic liquids could be useful in industrial applications, but so far there has been limited success harnessing their potential as catalysts on a commercial scale. The ISOALKY process solution is successful on a commercial scale because the catalyst, a chloroaluminate ionic liquid with a proprietary composition, has finely tuned acidity to avoid undesirable cracking, isomerization, and heavy by-product formation, as well as properties that allow low inventory, low consumption, efficient separation, and easy on-site regeneration.

Chevron Corp., one of the worldâ€™s leading integrated energy companies and Honeywell UOPâ€™s alliance partner for ISOALKY technology, is converting the existing hydrofluoric acid alkylation unit at its Salt Lake City refinery into the first alkylation unit in the US based on ionic liquid technology. ISOALKY technology takes ionic liquids a step closer to becoming mainstream in industrial applications, and can have a significant effect on global production of clean fuels using a process with lower environmental impact.

511680 Organotemplate-Free Synthesis of Hierarchical Beta Zeolites

Abstract:

Zeolite beta is a large-pore, low-Al microporous material that features an intergrowth of two or more polymorphs comprising a three-dimensional, 12-membered ring channel system with great potential in refinery, petrochemical and chemical applications. This presentation focuses on a straightforward approach to synthesizing hierarchical beta zeolites by exposing microporous zeolites to hydrothermal conditions amenable for beta crystallization, which simultaneously creates hierarchical pore structure while maintaining crystallinity and framework integrity without the use of organic structure-directing agents (OSDA), pore-directing agents (PDA), or mesoporous templates. The obtained hierarchical beta zeolites possess enlarged mesopores and exhibit improved catalytic performance for the conversion of bulky molecules, and are promising to be used in a variety of applications wherein the presence of large mesopores are required to reduce diffusion limitation and coking formation in beta zeolites.

509784 The Future Is Here: Robotic Catalyst Removal

Abstract:

WorleyParsons has been developing new technology using vacuum catalyst removal from refinery and petrochemical reactors via a remote-operated, screw-propelled vehicle. The development of CAROL (Catalyst Removal Amphirol) is in response to the increasing requirement among major hazardous facility operators to reduce human exposure to inert confined space entry. Several case studies from the beta testing period are examined. The analysis includes the challenges associated with the current catalyst removal techniques and how CAROL has been shown to provide advantages.

To date, testing has included construction of a test vehicle that has been used in simulated reactors. Video of the test vehicle shows operating characteristics.

The current test plan projects initial field tests in industrial units between November 2017 and March 2018. Preliminary work has focused on testing in two plants. The presentation will include progress and test results.

The focus plants for the initial test include;

- A refinery in the USA Midwest requires catalyst change-out on its silica guard reactors every 6 months. Light crusting and the formation of a large shelf have been experienced during the execution of previous catalyst dumping from these vessels. Human entry is required under inert conditions to knock down, break apart, and vacuum the remaining catalyst. In a similar service in 2014, a person was killed in a reactor in Germany when a wall of catalyst fell on him. CAROL can be deployed during the dumping process to break apart problem areas and to maintain a level catalyst bed.
- An LNG plant in Australia has nine large, single-bed dehydration vessels that require catalyst changeout on an 18-month cycle. Because the vessels are short and wide, dumping is not effective. The current method for catalyst removal is vacuuming via human entry which takes a full shift to accomplish. With ambient temperatures exceeding 32°C (90°F), risk of fatigue and heat exhaustion limit the time any one person can spend in the vessel to less than one hour. It has been demonstrated that CAROL can achieve instantaneous catalyst removal rates comparable to that of a human whilst reducing the time spent by workers in the reactor and the number of inert entries required by more than 80%.

The number of fixed bed catalytic reactors in the global refining and petrochemical industry is estimated to exceed 58,000. Catalyst removal from each of these reactors pose a risk to safety and/or the environment regardless of whether dumping, vacuuming or water flooding is employed. CAROL provides an alternative option that minimises risk to human life and improves efficiency.

513636 Using the Industrial Internet of Things to Monitor Process Equipment for Health, Performance, and Energy Consumption

Abstract:

An overview of the Industrial Internet of Things (IIoT) and how it can be applied to monitor industrial equipment for equipment health, performance, and energy consumption will be presented. It will include information on how to get started with a project, a roadmap to build deeper projects, and information on how to estimate the ROI for such projects.

511860 Using Automation Modernization As a Lever for Business Success

Abstract:

When your facility is challenged to improve productivity, process availability, and enterprise profits, modernization of your control system may provide the answer. By taking advantage of today's automation systems' advances, you can create more profitable operations and make a lasting difference. By positioning and employing modernization in the right way, management will see that these projects offer the operational benefits and return on capital that they are looking for, and they will more readily approve the project. They are also more likely to view the project team as company leaders who help the enterprise cost-effectively meet market needs. In this presentation, we will review the three key best practices that will ensure success in your modernization project and for your business.

512093 Advanced Process Modelling for Refinery-Petrochemical Integration

Abstract:

Oil refineries focus primarily on producing transport fuels. Feedstocks that flow to petrochemical plants from refineries tend to be consequential. There is now an accelerating trend to reduced dependency on transport fuels and this coincides with an increased demand for petrochemical derivatives. As a result, there is a growing effort to integrate refineries and petrochemical plants, so that refinery products are more aligned with usable petrochemical plant feedstocks and overall economics and environmental performance improve.

There are integration opportunities available today with minimal change to current refinery operating conditions. Refinery streams such as fuel gas, LPG, FCC light ends and naphtha can be fed to petrochemical plants and petrochemical plant streams such as hydrogen and C4s can be fed back to refineries. Energy, in terms of steam and electricity, can also be easily moved between the two, providing significant financial benefit.

The traditional route to generating lighter components from crude oil in a refinery is by feeding VGO or Vacuum/Atmospheric Resid to an FCC unit to generate olefins and aromatics. Extending this technology, crude oil can be cracked directly or the crude can be distilled first and the products cracked via steam or catalytic methods, or even both.

Refinery/petrochemical integration tends to increase the overall capital investment but the improved generation of higher-value products and the reduction in previously â€˜strandedâ€™ products makes integration financially very attractive.

When assessing new design or revamp, one significant challenge is how to adequately evaluate the costs and environmental benefits of refinery/petrochemical integration. The standard approach is to use Linear Programming models with simplified plant feed/product/cost models. However, this approach is fundamentally flawed because although it accounts for flexibility in stream routing, it does not take account of reactor catalyst performance in the units and the wide range of operating conditions across the refinery that are possible. A suitable overall process model is needed where stream routing, catalyst performance and operating conditions can all be simultaneously manipulated to achieve maximum profit. The associated model complexity is high yet it is essential that this be considered, otherwise significant potential benefit may be left unrealized. It could also mean that a profitable project is stopped due to an inappropriately chosen set of operating parameters.

This paper shows a new approach, where advanced process modelling techniques are used to tackle such complex problems and deliver insights to design parameters and stream allocation. The approach combines equation-based models with advanced model-pruning and equation solving techniques. These are used to determine optimum stream routing, plant configurations and operating conditions. A case study shows the benefits of the new approach. As a result, new or revamp refinery/petrochemical integrated designs are generated which fully exploit the operating freedoms available and deliver maximum profit.

513842 Re-Thinking Benzene Control Strategies to Maximize Octane

Abstract:

Limiting production by precursor removal from naphtha reformer feed or post reformer segregation and destruction by saturation are strategies that have been widely deployed to meet gasoline benzene limits required by EPA's MSAT 2 rule. These strategies not only reduce pool octane by limiting benzene but also tend to restrict naphtha reformer severity which in turn reduces availability of higher carbon number aromatics.

Vehicle manufacturers are zeroing in on higher octane for higher compression engines as a primary compliance strategy to meet fuel efficiency standards. Studies have shown that meeting the higher anticipated octane levels requires more octane boost than can be achieved by light naphtha isomerization and/or C4 alkylation. Higher severity naphtha reforming is the pathway for refiners to deliver higher octane but before that can happen the reformer severity constraint imposed by MSAT2 benzene limits must be removed.

This paper will discuss alternative processing strategies including benzene recovery by solvent extraction and propylene alkylation of benzene that can be deployed for benzene compliance as well as to meet higher octane for improved fuel economy.

513873 Development of a Two-Stage Thermal Deoxygenation Process for Fuels and Chemicals from Biomass

Abstract:

Thermal DeOxygenation (TDO) is the thermal decomposition of mixed salts of levulinic and formic acids that can produce hydrocarbon oils at yields approaching 80% of theoretical. These acids can be sustainably produced in stoichiometric quantities by acid hydrolysis and dehydration (AHDH) of cellulose. The TDO reaction was scaled up to a 50 L semi-batch reactor to produce more than 20 L of oil for fractionation, blending and hydrotreating. The products were thus demonstrated as suitable blendstocks for gasoline, diesel, jet fuel, and fuel oils. While scale-up facilitated kilogram quantities of products for testing, the carbon loss to char increased from 8.5% at the bench scale to 48% at 50 L floor scale. Isothermal reaction experiments have identified two reaction regimes: 1) a low temperature addition-condensation regime that is diffusion-controlled and 2) a high temperature ketonic decarboxylation regime that is reaction-controlled. Through this understanding, a two stage reaction process has achieved yields in the floor scale reactor similar to early bench scale results. Current process development efforts involve testing equipment to move from semi-batch to continuous operations and scaling up the TDO process to match the output of UMaineâ€™s one-ton-per-day AHDH pilot plant. The combined AHDH-TDO process has the potential to produce a suite of bio-based chemicals in addition to fuels, so additional results include techno-economic analyses which consider the potential for co-product revenue enhancement of renewable fuels.

513289 Development of Green Pathways for the Extraction of Medically Active Components Form Grape Waste

Abstract:

Medical sector growth continues to advance thanks in part to a boom in the biomedical industry. Increased technological advancement of this level has afforded people health and wellness not seen in previous generations leading to longer more productive lives. Growth in the Biomedical can be amenably paired with green technologies to further the field in an environmentally friendly way. Innovation in the use of natural components in health fields as a way of improving health is a most promising direction but is one existing with many bottlenecks. Despite the challenges, these technology shifts are necessary to stay in line with environmental regulation, FDA regulation, and public opinion.

The use of Supercritical CO_2 is growing more and more common in many fields. It can be used singularly, with co-solvent, or as part of more complex systems to process Biomass in a more environmentally friendly way. Waste biomass can be especially useful here. What was once not worth processing and was relegated trash or burning can be processed to obtain high valuable compounds. Grape waste from wineries and juicing facilities consists of seeds, stems, rings, meat, and leaves. Most of this waste has little to no value and is relegated to trash or animal feed. With some creating processing, this waste can be a source of valuable oil as well as high-value antioxidants, polyphenols, and flavanoids. Scale is important in these operations and thus a pilot study is being conducted on this process to find the range of capacities that are profitable for this endeavor.

512995 Catalytic Option(s) for Hydrotreaters Struggling with Cracked-Rich Feeds

Abstract:

To help refiners satisfy today's fuel quality specifications while hydroprocessing cracked-rich heavy distillates, development of step-out next generation hydroprocessing catalysts is needed. The application of such catalysts to the hydroprocessing of cracked-rich, heavy distillate blends should help to achieve the desired product qualities, such as diesel cetane, specific gravity, and/or product yields. Additionally, placement of the catalyst bed containing the new catalyst relative to conventional, supported hydrotreating catalysts should likewise be applicable to identify more advantageous options for the relief of processing constraints. This presentation will include reviews of these examples from pilot plant studies, highlighting the apparent advantages of at least one step-out, next generation hydroprocessing catalyst. This presentation will also cover at least some of the limitations of such a catalyst.

513669 Smart Equipment Integration for Improved Process Yield

Abstract:

A process design is presented where integration of key equipment within a hydrogen production unit produces increased process benefit. A utility consuming process in turn becomes a utility producer. Through this same flow sheet revision a problematic gas stream is re-routed to equipment that is better suited for its usage, emissions are potentially reduced, and process yield is increased.

Innovative ZoneFlow® Reactor Technology- Gearing for Future in Steam Reforming for Hydrogen Generation

Sanjiv Ratan, ZoneFlow Reactor Technologies LLC, USA
Prof. Juray De Wilde, Universite Catholique de Louvain, Belgium

Introduction

The predominant technology for hydrogen generation over the past decades has been steam reforming of light hydrocarbons, mainly natural gas commonly called steam-methane reforming (SMR). owing to its favorable attributes This mature technology, however, has seen only incremental improvements in its catalyst design and its performance in conjunction with superior tube metallurgy and reformer design with the aim for better efficiency and reliability.

ZoneFlow™ Reactor Technologies LLC (ZFRT) has developed proprietary, radically innovative structured catalyst solutions for steam reforming, pre-reforming and post-reforming. ZoneFlow (ZF) reactors provide higher heat transfer, lower pressure drop, higher geometric surface area (GSA), more uniform flow and heat distribution, and long-term structural integrity compared to the current state-of-the-art pellet catalyst. The use of ZF reactors enables significant capital and operating cost reduction in steam reforming both for new plants as well as revamps.

Current (best available) technology

The conventional "pellet" steam reforming catalysts suffer from few inherent limitations and/or operational deficiencies. These mainly relate to a) uneven random packing voidage leading to undesired flow and temperature mal-distribution, b) catalyst attrition and breakage from thermal cycling leading to increasing pressure drop and related capacity limitations and c) pore-diffusion limited reaction path leading to curtailed intrinsic activity and lower resistance to carbon formation.

With the steam reformer being the heart of a hydrogen plant carrying out the complex steam reforming reaction involving heat, mass and momentum transfer, its reforming catalyst plays a critically important role in its thermal design, performance, reliability and cost-effectiveness.

Past attempts to develop structured steam reforming catalysts to overcome some of these limitations have suffered from the innate problems around differential expansion gaps leading to feed by-passing and disturbed boundary layer endotherm, as well as from the reaction hydraulics being away from the tube wall, lacking required heat input, thus leading to insufficient conversion and hotter tubes.

Innovative Breakthrough

ZF Reactor technology not only overcomes the above-mentioned deficiencies but also offers an unmatchable combination of higher heat transfer with substantially lower pressure drop. The ZF Reactor's unique structure also imparts a high degree of proximal flexibility which eliminates the differential thermal expansion gaps, while carrying high (metal substrate) physical strength against breakage over its lifetime.

Its nested modules provide much more geometric surface than pellets, and take away the diffusion-limited access to the layered active sites thus resulting in much higher intrinsic activity and increased resistance to carbon formation compared to pellets.

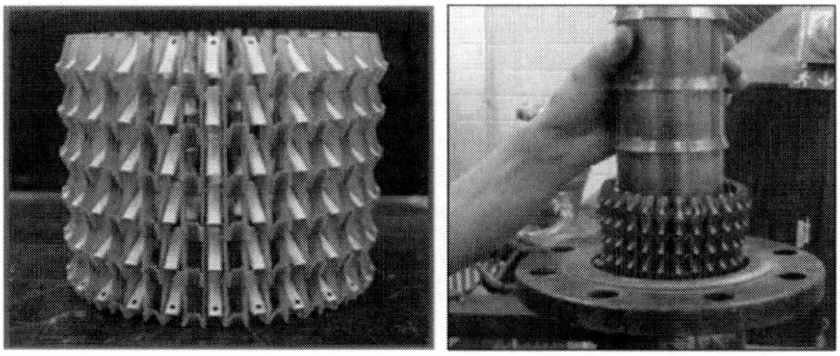

Figure-1 : ZF Reactor's typical casing and assembly

Comparative Merits of ZF Reactors vs. Conventional 'Pellet' Catalyst

ZF Reactors are advanced engineered catalyst systems based on unique 'annular' foil casing structure (see Figure-1) offering two primary advantages over conventional 'pellet' steam reforming catalysts :

- Lower pressure drop (dP) – the ZF uses dP mainly to enhance the turbulence near the tube wall, avoiding to large extent the "unutilized" dP in overcoming flow and diffusion resistance typical of random pellet packed column.
- Superior heat transfer – ZF annular structure directs the gas flow to impinge the hot surface of the tube wall which significantly enhances heat transfer and also lowers the tube skin temperatures, thus also permitting process intensification

Additionally, ZF reactors offer following significant advantages compared to pellets for revamping as well as new steam reformers.

- Much higher catalyst effectiveness for driving the endothermic steam reforming
- Uniform heat transfer and fluid flow along the tube length as well as across the tube rows; also eliminating hot spots that can restrict full-load operation and risk shorter tube life
- Long-term structural integrity, including mechanical features to accommodate thermal cycling, thus avoiding pressure drop build-up over time related to progressive deterioration of pellets

ZF development and validation milestones

Various Computational Fluid Dynamics (CFD) and Finite Element Analysis (FEA) studies have not only confirmed these properties but also allowed gaining deeper insights into the ZF design, also for optimizing the balance between these key performance parameters. ZF Reactors can be tailored to application-specific optimum combination of higher heat transfer coefficient and lower dP compared to conventional pellet catalysts as shown in Figure-2

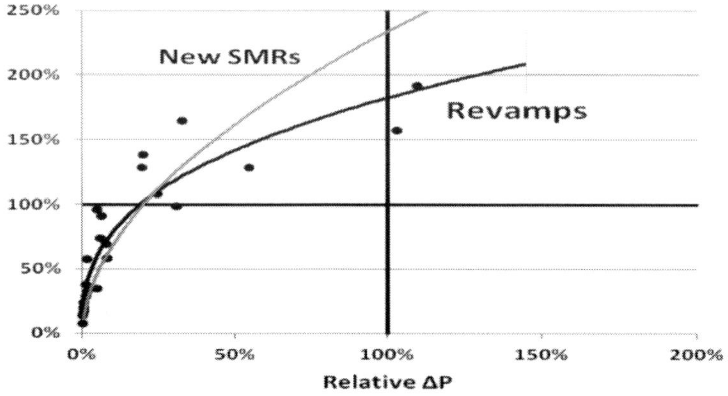

Figure-2 : Optimization of HTC against Pressure drop

ZFRT has achieved various development and validation milestones over recent years:

- Bench-scale validation at the University of California at Davis HyPaul labs.
- Independent CFD kinetic modeling for improvements and/or optimization of simultaneous heat and mass transfer against pressure drop. (Figure-3)
- FEA to establish mechanical robustness and long-term durability (Figure-4)

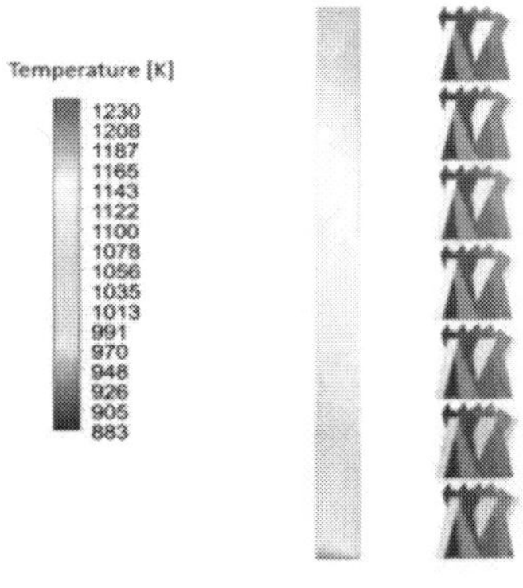

Figure-3: CFD modeling

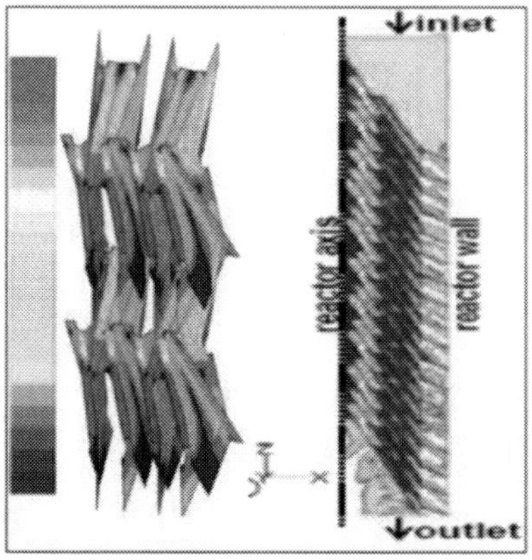

Figure-4: ZF FE Analysis Results

- Demonstration in a large commercial SMR with ZF put in 2 tubes for close to 2 years, allowing direct comparative assessment conventional BAT pellet catalyst under the same operating conditions, apart from ZF loading / unloading verification.
- Nearing completion of its own designed and built pilot plant installed within the Université Catholique de Louvain, Belgium (UCL) to further validate the performance and physical integrity of the ZF Reactor designs under wide range of conditions.

- ZF's performance parameters have been validated through advanced CFD analysis, followed by the upcoming pilot plant testing under various commercial operating conditions. ZF designs have been further tested in a commercial SMR (in 2 tubes out of more than 200) validating near 2-fold higher heat transfer together with significantly lower pressure drop compared to current pellet steam reforming catalysts.

- The ZF Reactor designs can be optimized and customized for required balance between heat transfer and pressure drop for the specific application, and can also be made to fit any commercial SMR tube size.

Added-Value ZF Solutions for Hydrogen SMRs

Following case analysis are presented covering various ZF applications :

- **ZF-Single pass (SP)**
  - De-stressing and Debottlenecking SMRs
  - Higher-flux, cost-effective and more reliable new SMRs
- **ZF-Convective pre-reforming (CPR)**
  - Unmatched in-situ retrofit for additional capacity in existing SMRs without major modifications
  - Efficient and cost-effective applications in new SMRs
- **ZF-Bayonet (B)**
  - Uniquely applicable for recuperative reforming, overcoming current barriers

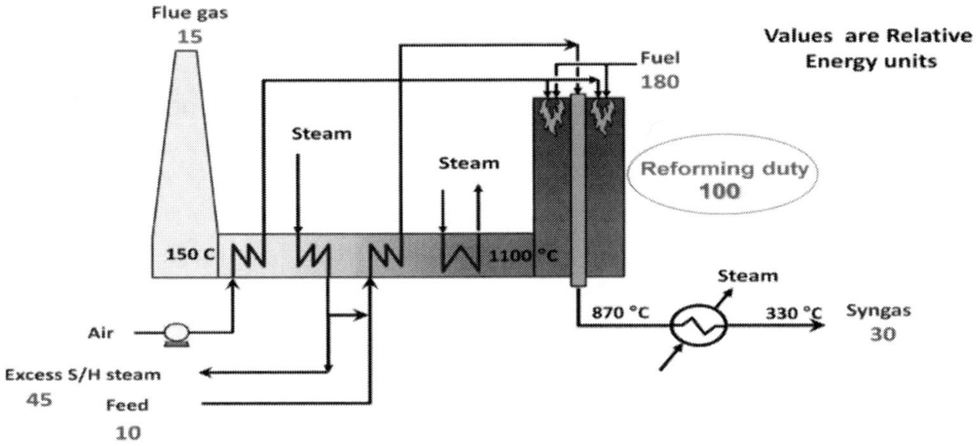

Figure-5 : Steam-methane Reformer (SMR) : Heart of a Hydrogen Plant

The common underlying advantages from ZF's lower dP and higher heat transfer coefficient include:

- higher outlet temp without increasing max. tube skin temperature (TSM)
- higher heat flux and/or higher outlet temp without increasing bridge-wall temperature, or related SMR firing duty
- lower Approach to Equilibrium

Case 1 Stressed-SMR attributes and make-over using ZF-SP reactors

- Pressure drop (or its build up) limiting throughput to under-nameplate capacities
- Loss of catalyst activity → hotter tubes limiting outlet temperature
- Catalyst attrition / settling from thermal cycling
- Carbon formation / hot spots
- Impact on (remaining) tube life
- Constrained reforming capacity
- Replacing (pellet) catalyst in these SMRs with ZF-SP Reactors can overcome these deficiencies, as summarized in following case analysis :

Table-1: SMR De-Stressing	SMR Design	Stressed Operation	ZF-SP replacing pellets
Relative Capacity, %	100	95	100
Capacity limitations	-	dP, TSM	removed
S/C Ratio	3.0	3.3	2.8
Outlet temp, C	860	843	868
Approach to Equilibrium EOR C	-10	-12	-7
CH4 slip, dry vol %	5	5	5
Radiant Pressure drop, bar	2.8	2.8	2.3
Relative Radiant duty %	100	97	99
Avg heat flux kW/m2	78	74	78
Bridgwall temp, C	1000	980	998
TSM (avg) C	935	935	926

Case 2 ZF-SP Reactors for SMR Debottlenecking

- Achieve capacity in excess of nameplate (~ 105 %), while utilizing design margins in SMR burners and fans
- Higher average heat flux without exceeding tube design temperature
- Improved temperature uniformity
- Extended tube life and improved reliability
- Better catalyst performance and "life cycle" costs
- Optimized operation and enhanced reliability

Table-2 : SMR Debottlenecking	Reference	ZF-Radiant
Relative Capacity, %	100	105
Capacity limitations	dP, TSM	removed
S/C Ratio	3.0	2.8
Outlet temp, C	865	870
Approach to Equilibrium C	-10	-7
CH4 slip, vol %	5.5	5.7
Radiant Pressure drop, bar	2.8	2.5
Relative Radiant duty %	100	104
Avg heat flux kW/m2	75	78
Bridgwall temp, C	**1008**	**1004**
TSM C	**940**	**938**

Case 3 ZoneFlow™ Convective Pre-reforming (ZF-CPR)

ZFRT has also designed and engineered an ultra-low pressure drop structured catalytic reactor with high surface area / low cross section for non-adiabatic low-temperature convective pre-reforming (CPR).

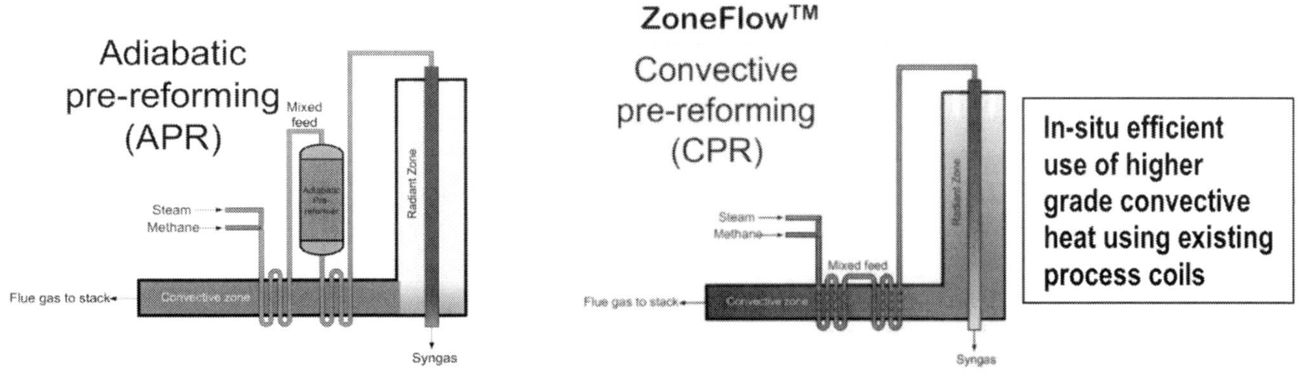

- In-situ horizontal loading in the mixed feed preheat (MFPH) coil in the convection
- Optimization of pressure drop in combination with ZF-SP in SMR radiant tubes
- For revamps, 8-12% additional reforming without increasing SMR firing duty
- For New SMRs, 10-15% smaller SMR and proportionately also the steam system.
- No concerns over horizontal loading catalyst settling, attrition and related feed by-passing
- Shifts steam loads to Utilities and Offsites , thus also lowering core-plant C-footprint

Table 3: ZF-CPR Integration		Existing	ZF + CPR
Relative Capacity,	**%**	**100**	**115**
S/C Ratio		**3.0**	**2.8**
SMR inlet temp,	C	550	550 / 575 *
SMR outlet temp,	C	870	870 / **878** *
Approach to Equilibrium EOR	C	-10	-7
CH4 slip,	dry vol %	4.8	5.4 / 5.0
Radiant Pressure drop,	**bar**	**2.5**	**2.3**
Avg heat flux	kW/m2	80	80 - 82
Relative Radiant duty	**%**	**100**	**100 / 103**
Bridgewall temp,	**C**	**1020**	**1009 / 1018**
TSM	**C**	**950**	**938 / 948**

ZF-Bayonet (B) for Recuperative Reforming

ZF Reactors are inherently suited for "bayonet" (tube-in-tube) recuperative reforming based on their annular casing design. The reformed process gas flows out through the center (core) tube providing part of the high grade heat recovery for steam reforming , thus lowering the fired duty as well as size of the reformer by 15-20% . ZF also overcomes the main risk of current "pellet" catalyst crushing from differential thermal expansion as an unmatched technology based on its robust metallic substrate and flexible casing structure.

.

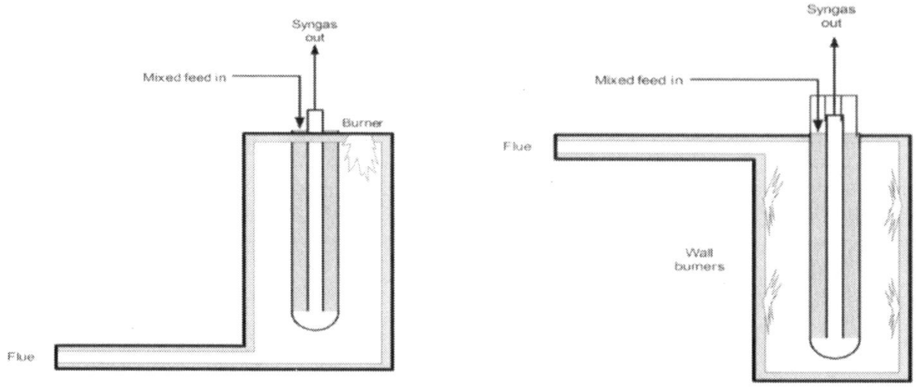

Economic Benefits and Applications

These ZF Reactor merits can translate into attractive benefits ranging from relieving "stressed" reformers, extending reformer tube life and capacity revamps to lowering SMR related OPEX and CAPEX. It covers the capacity span from large hydrogen generation plants for the refining sector (whether new or revamp of existing plants) to the advanced solutions for small scale distributed hydrogen plants for fuel cells and futuristic hydrogen energy landscape. It's advanced merits also cover the related Syngas production for methanol and GTL plants, including mini modularized units for remote gas monetization as well as Flare gas reduction.

Based on the energy costs and economic investment / IRR criteria for new plants and the level of modifications in revamping of existing plants, each application needs to be evaluated on a case-to-case basis. in terms of Opex and Capex analysis to select the most appropriate, conforming and economically attractive (ZF) solution. Other important case-specific considerations may include stand-alone units with no export steam, lowering of carbon footprint, SMR intensification, long term feed & product pricing, revamp project downtime as well as environmental aspects for overall optimization of hydrogen generation.

Conclusions

ZoneFlow^{TM} Reactor Technology (ZFRT) developed innovative, and differentiating structured catalyst technologies with demonstrated improvement in heat transfer, pressure drop and catalyst effectiveness. Further , the new pilot plant designed and constructed at UCL, Belgium and also the collaborative efforts with some of the prominent players in the syngas field, will allow further rigorous testing of ZF reactors under commercial conditions for application in steam methane reforming. The portfolio also includes pre-reforming and recuperative reforming. ZF reactor technologies offer attractive applications and promising solutions in steam reforming and for realizing advanced options, thus facilitating cost-effective , efficient, reliable and distinctly attractive hydrogen generation.

513253 Catalyst Characterization in the Molecular-Level Kinetic Modeling of Hydroprocessing Reactors

Abstract:

The commercial upgrading of conventional and unconventional hydroprocessing feeds to valuable petrochemicals and fuels usually involves a series of reactors, each of which can contain multiple beds with several different layers of catalysts. A multiphase hydroprocessing reactor system with molecular detail of the feed and products was modeled using an in-house software, the Kinetic Modeler's Toolbox. A reaction network of hydroprocessing feeds with $O(10^3)$ species and $O(10^4)$ reactions that includes the detailed hydroprocessing chemistry (saturation, hydrocracking, ring opening, hydrodesulfurization, hydrodenitrogenation, and hydrodeoxygenation) was generated. Several coke formation pathways and metal deposition reactions were also modeled to account for time-on-stream catalyst site reduction. The initial flow rates of the feed were calculated by defining an attribute, or structural moiety, probability density function tree and optimizing the difference between the feed experimental and simulated properties. A set of catalytic rate law based material balances, one for each species, and an energy balance along with the feed flow rates was used to define the initial value problem to be solved in the kinetic model. Each reactor bed was modeled as a series of plug-flow pseudo-reactors with side-by-side reaction and vapor-liquid equilibrium, where each pseudo-reactor represents a separate catalyst layer. The mixing of the hydrogen quench streams between beds was modeled with a mass balance and an enthalpy balance. Additionally, a complementary selectivity-activity modifier model was used to model the catalyst deactivation on the longer timescale of the catalyst life. Linear free-energy relationships (LFERs) and quantitative structure-property relationships (QSPRs) were used to reduce and tune the parameters of the LHHW rate equations. The LFER concept was extended to catalyst families in order to further minimize the number of parameters and data needed for different catalysts[1]. The final model represents a real refinery reactor simulation over a range of process conditions like space velocity, temperature, pressure, and hydrogen flow. It also allows for scenario studies to test catalyst deactivation for different feed compositions and the impact of changing catalyst layer order, bed height, and quench rates.

[1] Horton, S.R., Klein, M.T. "Reaction and Catalyst Families in the Modeling of Coal and Biomass Hydroprocessing Kinetics." *Energy & Fuels*, 28:1 (2014), 37–40.

512390 Molecular Kinetic Modeling in Hydrocracking

Abstract:

Hydrocracking is in an exceedingly competitive commercial environment which necessitates licensors to provide increasingly sophisticated technical solutions. In recent years, hydrocracking solutions have expanded to catalyst stacking options, configuration studies and flexible product objectives. Next generation discoveries will depend on the ability to understand how different molecules interact with different catalysts and materials.

Historic limitations in analytical capabilities have driven most to implement lumped kinetic models where the molecules are grouped by their physical properties such as boiling point. While this approach enables accurate estimates of product yields within the scope of experimental data used to generate the model, it inevitably obscures molecular information due to multicomponent nature of each lump rendering the model without predictive capability for variable feeds, process configurations and materials.

While molecular kinetic models offer the potential to be predictive, they are immensely difficult to construct due to the vast number of species, reactions and associated rate constants. Recent advancements in analytical capabilities to enhance feed and product characterization as well as developments of mathematical methods to reduce the number of rate constants have made it possible to attempt feedstock-independent models that care applicable in around the commercially relevant feed and process conditions.

In this talk, hydrocracking molecular kinetic modeling and itâ€™s applications will be presented. Case studies will be shown to demonstrate the capabilities of the model in development of novel technical solutions.

513618 Lumped Kinetic Model for Hydrocracking of Heavy Oils over Oil-Dispersed Mo-Based Catalyst

Abstract:

A 5-lump (unconverted residue, vacuum gas oil, distillates, naphtha and gases) kinetic model for hydrocracking of heavy oils is developed. To obtain the experimental data, a continuous-stirred tank reactor has been used and the experiments have been carried out using Mo-based oil-dispersed catalyst precursor and vacuum residue from the commercial refinery in South Korea. The reaction temperature has varied in the range of 425 â€" 440 Â°C and the reactor pressure has varied in the range of 140 â€" 160 bar, which are the ordinary operating window of the slurry hydrocrackers. For the efficient parameter estimation, the efficient optimization algorithm combining the genetic algorithm and the deterministic algorithm have been adopted due to the nonlinear and high dimensional object function for the parameter estimation. The prediction results by the developed kinetic model has shown the good agreement with the experimental values so this work is considered to contribute to designing an appropriate slurry hydrocracker.

512904 Change of Product Yield and Quality with a Variation of Residue Conversion in a Slurry-Phase Hydrocracking of Vacuum Residue

Abstract:

Slurry-phase hydrocracking (SPH) technology is one of the promising technologies for upgrading of heavy oils in the refining area. Since it uses a nano-sized dispersed catalysts the asphaltene, which is known for the difficult component to be treated by heterogeneous catalysts, can be hydrogenated and converted into a light molecular sized component. However, SPH is still on-going demonstration test in a commercial field and it is not found from the literature what characteristics SPH has in a once-through operation mode. In this respects, this study investigated the change of product yield and quality with a variation of residue conversion through a once-through mode in a slurry-phase hydrocracking of vacuum residue.

The experiment was carried out with a function of reaction temperature from 420 to 440 â„ƒ and liquid hourly space velocity (LHSV, 0.3 â€“ 0.5 1/h), hydrogen partial pressure (80-160 bar), H_2 to feed ratio (600-2000 Nm^3/kl) with molybdenum catalyst concentration of 500 wt.ppm based on metal content on feed in a 4L slurry bubble column reactor. For the analysis of the product yield and properties, gas composition was determined using a gas chromatograph (Agilent 7890) while the liquid properties for the API (D 5002-99), sulfur and metals (nickel and vanadium, D4294-03), micro-carbon residue (MCR, D4530-06), asphaltene (normal heptane insoluble content, D2007), and boiling point distribution (D7500) were all analyzed according to standard ASTM methods. Additional, the amount of sediments and the remained catalyst concentration in the liquid product was examined after reaction.

TUNING PID CONTROLLERS FOR COMPLEX INTEGRATING PROCESS DYNAMICS

James Beall

Emerson Automation Solutions, Austin, TX

Biography

James Beall is a Principal Process Control Consultant at Emerson Automation Solutions with over 36 years of experience in process control. He graduated from Texas A&M University with a Bachelor of Science in Electrical Engineering in 1981 and worked for Eastman Chemical Company from 1981 until 2001 at which time he joined Emerson. His areas of expertise include process instrumentation, control valve performance, control strategy analysis and design, advanced regulatory control and multivariable, model predictive control. He has designed and implemented process control improvement projects in the chemical, refinery, pulp and paper, power, pipeline, gas and oil, and pharmaceutical industries.

Paper ID 513646

Prepared for Presentation at American Institute of

Chemical Engineers 2018 Spring Meeting

14th Global Congress on Process Safety

Orlando, FL April 22-26, 2018

Keywords: Control, PID, tuning, integrating, integrator, lag, lambda.

Abstract

Tuning PID controllers for integrating processes, such as levels, can be frustrating and counterintuitive. For example, reducing the controller proportional gain on an oscillating loop can increase the oscillation amplitude! In addition to the basic integrating response, two other more complex integrating type responses are often encountered in many manufacturing processes. These are "integrator plus dead time" and "integrator plus dead time and lag". Examples of a "dead time plus integrator plus lag" response include boiler drum level, de-aerator level, large volume reactor temperature, and large volume gas pressure. This paper will show how to recognize these types of integrating responses, measure the process dynamics and provide a straightforward method to calculate PID tuning parameters. Most importantly, this method can be used to calculate controller tuning to achieve the control response that optimizes the process performance. For example, in some cases the best process performance requires aggressive tuning with good load regulation capability. In another case it might require the "least aggressive" tuning, that maintains a process variable (e.g. level) within the maximum allowable deviation from set point, yet provides maximum attenuation of process variability. This tuning method can also be applied to lag dominant or "near integrating" processes which greatly reduces the time to measure process dynamics and can provide superior load regulation. Plant examples will be used to illustrate the concepts and results.

Introduction

The two most common categories of process responses in industrial manufacturing processes are self-regulating and integrating. A self-regulating response is characterized by a response of the process variable that stabilizes, or "self-regulates", to a new value sometime after the input to the process is moved to a new value. From the standpoint of a "proportional, integral, derivative" (PID) process controller, the output of the PID controller is the input stimulus to the process and the process variable is the input to the PID controller, also called the process variable. Figure 1 compares the response of the process variable to a step change of the PID controller output for a self-regulating process and for an integrating response. Both responses include dead time. Note that the self-regulating process PV stabilizes or "self-regulates" to a new steady state value while the PV of the integrating process continues to change. The response of the latter is an "integration" of the difference between the process inputs and its outputs, hence the term "integrating process".

Level processes typically have an integrating response, the likely exception being when the outflow of the vessel is gravity driven. Other processes can have an integrating response. A "low pressure, large volume" gas pressure control application typically has an integrating process. Another example of an integrating process is a reactor temperature controller that cascades to a "jacket temperature difference" controller. The "jacket water inlet temperature difference" controller controls the difference between the reactor contents temperature and the jacket water inlet temperature based on the set point specified by the output of the reactor contents temperature controller. Figure 2 shows these examples of integrating processes.

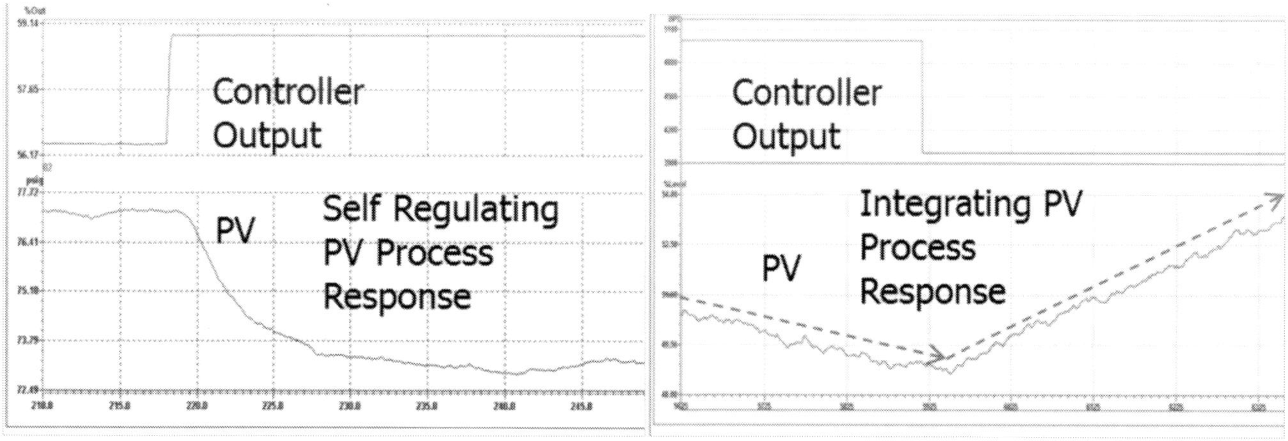

Figure 1 Response of the PV to a Step Change of the Controller Output for a Self-Regulating and an Integrating Process

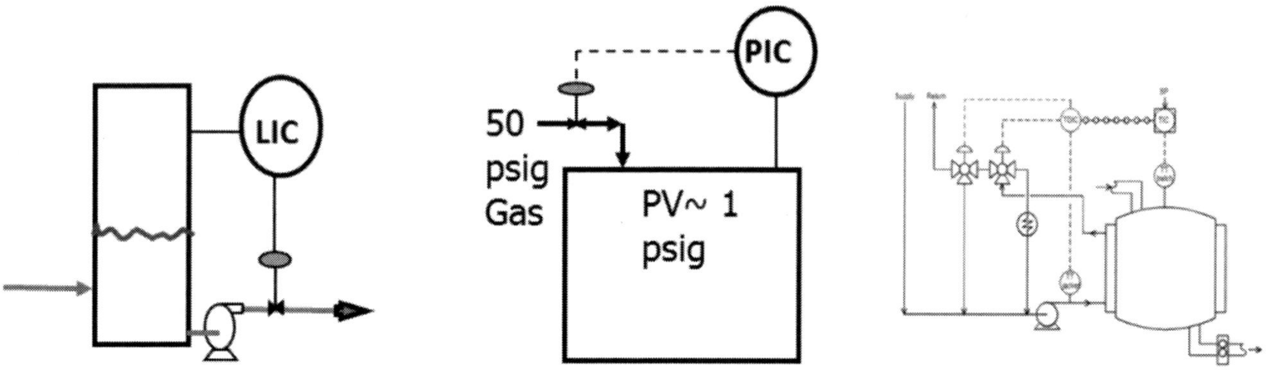

Figure 2. Three examples of integrating processes, Level, Gas Pressure and Temperature

Challenges

One of the challenges of tuning a PID controller for an integrating process is that when the integral action of the controller is combined with the integrator function of the process, the control loop will oscillate if the integral action of the controller is "too fast" (i.e. the integral time is too short). It is not intuitive to know when the integral time is too short. Another challenge is that most PID tuning methods for integrating processes are very aggressive and do not provide a method to adjust the aggressiveness of the closed loop response. If the controller proportional gain (P) is reduced to make the closed loop response less aggressive, the loop is more likely to oscillate! This is quite the opposite result than when this tactic is used on a self-regulating process! Furthermore, there is significant financial benefit by tuning control loops for desired closed loop response to maximize process performance.

Discussion

There are several variations of the integrating response that are commonly found in industrial processes. The 3 most common variations of the integrating response include the following and they will be analyzed in this paper. The step response of each of these types of dynamics is shown in Figure 3.

- Integrating
- Integrating plus dead time
- Integrating plus dead time plus lag

For the common "liquid in, liquid out" level process, the process dynamics typically do not include significant dead time. However, if the level process involves a transport delay of solids material to a bin, the response could include dead time. Boiler drum levels often have an apparent dead time caused by the inverse response of boiler feed water. A dead time could be introduced if the process measure is delayed by the sensor location, for example a temperature measurement in the outflow line of a reactor. If the level controller is cascaded to a flow loop, the closed loop response of the flow loop is included in the level response which would likely result in a lag (small if the flow loop is tuned properly!). The integrating process might include a lag such as that exhibited by a large stirred reactor temperature or a boiler feed water deaerator tank level. These types of complex integrating responses are quite common in industrial processes across all industries.

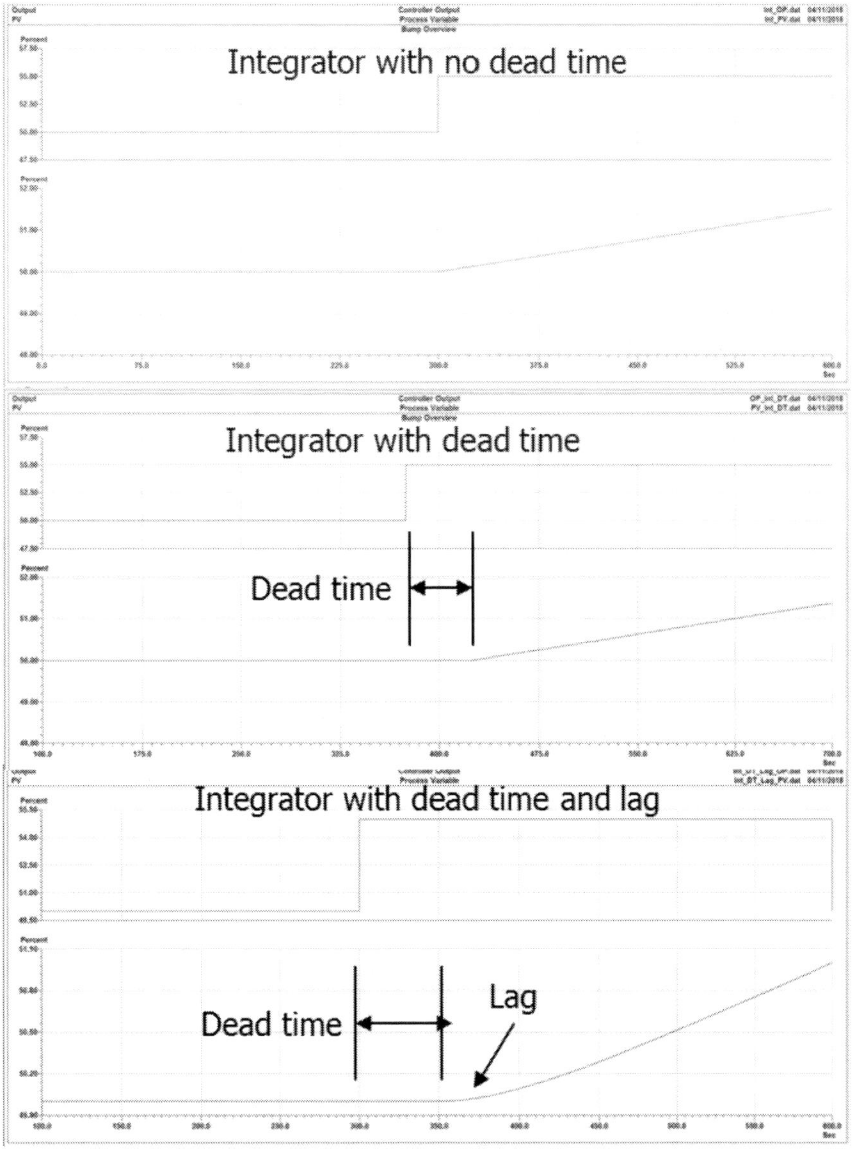

Figure 3 Step Response from Integrating Process with Different Dynamics

The integrating level process with an outlet flow disturbance shown in Figure 4 will be used for the tuning calculations presented in this paper. The 3 most common integrating process responses will be evaluated for this process.

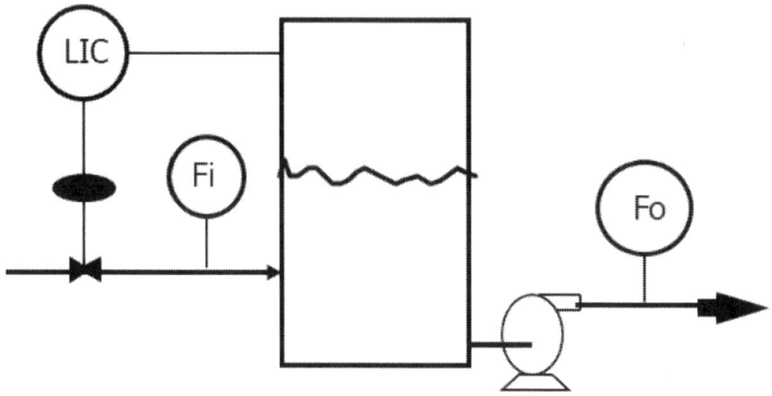

Figure 4 An Integrating Level Process With Load Disturbance Flow, Fo

Figure 5 shows the block diagram of a PID controller applied to an integrating process and a load disturbance, Fd, shown in Figure 4

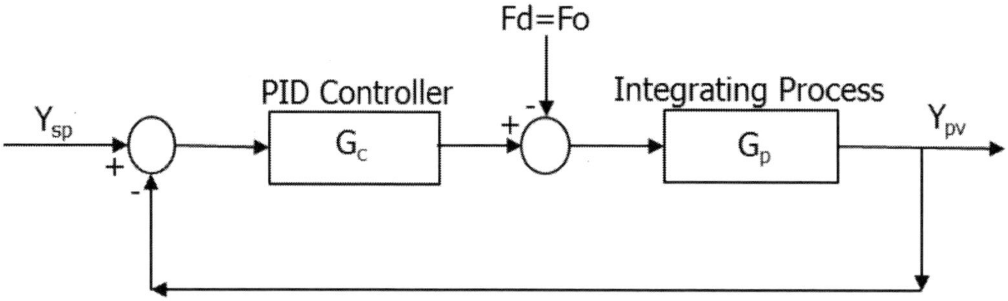

Figure 5 Block Diagram of Integrating Level Process With Load Disturbance Flow

Case 1: Integrating Process with a Proportional Action Only Controller

In case 1, consider an integrating response, without deadtime, and a PID controller with no integral or derivative action. The Laplace transform of the process is shown in equation 1. The Laplace Transform of the "proportional only" PID controller is shown in equation 2. The open loop transfer function, G_{lp}, for this case, is shown in equation 3.

$Gp(s) = \frac{Kp}{s}$ Kp has units of %PV/Second/%OUT Equation (1)

$G_C(s) = Kc$ Kc has units of %OUT/%PV Equation (2)

$Glp(s) = \frac{KcKp}{s}$ Equation (3)

The closed loop transfer function from SP to PV is

$Gsp\ (s) = \frac{\left(\frac{KcKp}{s}\right)}{1+\left(\frac{KcKp}{s}\right)}$ Equation (4)

$Gsp(s) = \frac{KcKp}{s+KcKp}$ Equation (5)

$Gsp(s) = \frac{1}{\left(\frac{1}{KcKp}\right)s+1}$ Equation (6)

$Gsp(s) = \frac{1}{\lambda s+1}$ Equation (7) where $\lambda = \frac{1}{KcKp}$ (units of seconds)

Thus, the closed loop setpoint response is a first order with closed loop time constant λ, lambda. Solving for the proportional only tuning constant, Gain (Kc) results in the following tuning rule.

$Kc = \frac{1}{\lambda Kp}$	Equation (8)

Using an example integrating process with a process gain of 0.001 %PV/Second/%OUT and with 0 dead time, the controller gain Kc for a chosen λ of 500 seconds would be as follows.

$$Kc = \frac{1}{500seconds * \left(0.001 \frac{\frac{\%PV}{Second}}{\%OUT}\right)} = 2.0\ \%OUT/\%PV$$

Figure 6 shows a simulation of the set point (SP) response of this process and a PID controller with a gain of 2.0 %OUT/%PV (no integral or derivative). Note the closed loop time constant of 500 seconds as chosen and 0% offset between the SP and PV at steady state. Figure 7 shows the simulated response to a step load disturbance of the outflow, Fo.

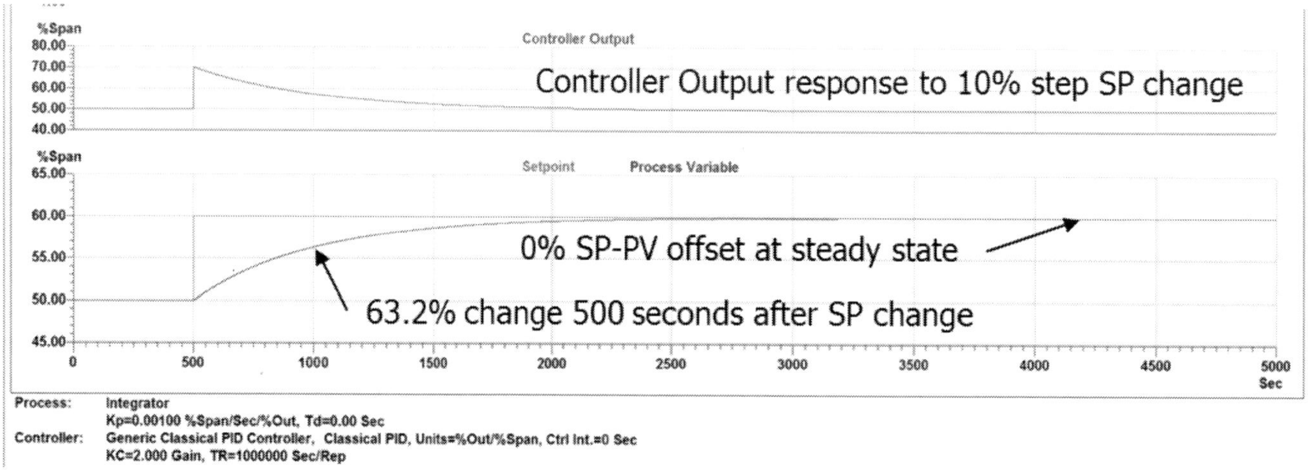

Figure 6 Simulation of Step SP Change for Integrating Process and Proportional-Only PID Controller Tuned for Closed Loop Time Constant, λ, of 500 Seconds

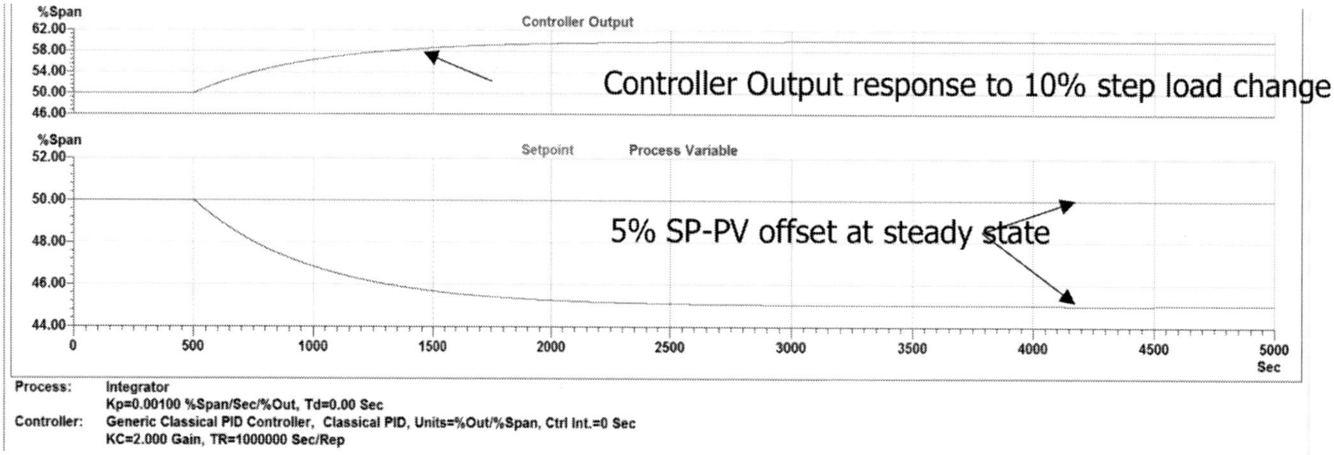

Figure 7 Simulation of Step Load Change for Integrating Process and Proportional-Only PID Controller Tuned for Closed Loop Time Constant, λ, of 500 Seconds

Case 2: Integrating Process with a PI Controller

Note that proportional-only control of a step load disturbance for an integrating process results in an offset between the SP and PV at steady state. This suggests the need for integral action in the controller to eliminate the SP-PV offset. However, if the integral action is too fast, the closed loop response will be oscillatory. The goal is to develop and equation for both the proportional gain and integral action that will provide a non-oscillatory response.

The transfer function, Gc, of a series PID controller with proportional and integral action is:

$$Gc(s) = \left[Kc * \left[1 + \frac{1}{TrS}\right]\right] \qquad \text{Equation (9)}$$

Where Kc has units of %OUT/%PV and Tr has units of Seconds.

The open loop transfer function for the integrating process and series PID controller with proportional and integral action as shown in Figure 5 is:

$$Glp(s) = Gc(s) * Gp(s0 = \left[Kc * \left[1 + \frac{1}{TrS}\right]\right] * \left[\frac{Kp}{S}\right] = \frac{Kc(TrS+1)}{TrS} * \frac{Kp}{S} \qquad \text{Equation (10)}$$

$$Glp(s) = \frac{KcKp(TrS+1)}{TrS^2} \qquad \text{Equation (11)}$$

The root locus for equation (10) and (11) is shown in figure 8. The root locus reveals that with a small reset time Tr (fast integral action) and a low controller gain Kc, the closed loop poles will be very lightly damped and the loop will be likely to oscillate. This is often the case in practice and the typical response to reduce the controller gain for an oscillating loop will make it oscillate even more! To avoid the oscillation due to controller tuning, the tuning should be chosen to locate the closed loop poles all the way around the circle back to the negative real axis where the damping coefficient $\zeta=1$.

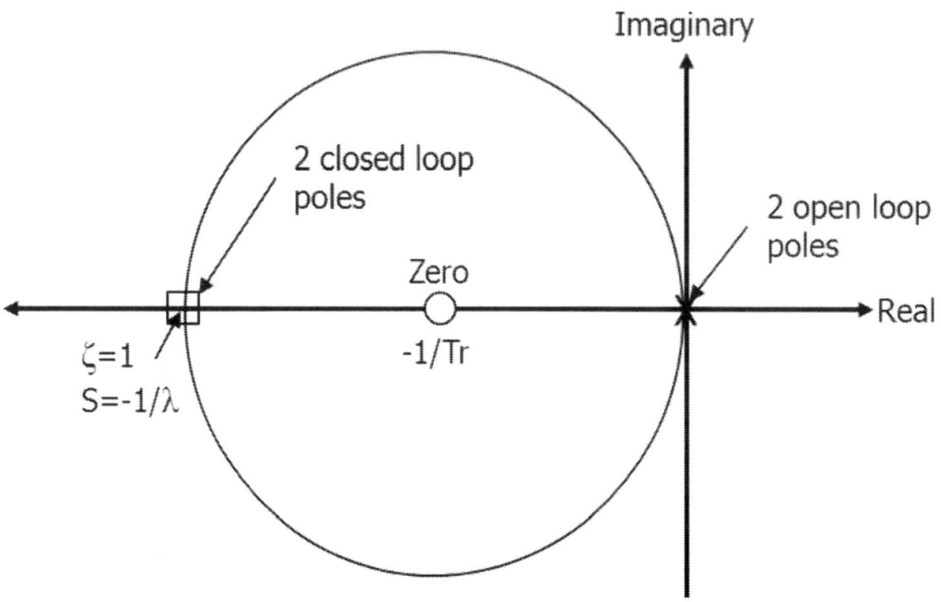

Figure 8 Root Locus diagram for the Open Loop Transfer Function of an Integrating Process and PID Controller with no Derivative Action

Referring to figure 5, the closed loop transfer function G_{ld} from the disturbance Fd=Fo to the process variable Y_{pv} for a series PID controller with proportional an integral action for the integrating process is:

$$Gld(s) = \frac{-Gp(s)}{1+Gc(s)*Gp(s)} = \frac{-\frac{Kp}{s}}{1+\frac{KcKp(TrS+1)}{TrS^2}} = -\frac{\frac{Tr}{KcS}}{\frac{Tr}{KcKp}S^2+TrS+1} \quad \text{Equation (12)}$$

Thus, the characteristic equation for the closed loop control system is:

$$0 = \left(\frac{Tr}{KcKp}S^2 + TrS + 1\right) \quad \text{Equation (13)}$$

The solution to the quadratic equation (13) are the closed loop poles. To have a critically damped closed loop response, the controller tuning parameters Kc and Tr are chosen to have two closed loop poles at -1/λ. When this occurs, the resulting characteristic equation must be as follows:

$$0 = (\lambda S + 1)^2 = \lambda^2 S^2 + 2\lambda S + 1 \quad \text{Equation (14)}$$

Comparing this to closed loop characteristic equation (13) results in the following equalities:

$\frac{Tr}{KcKp} = \lambda^2$ Equation (15)

$Tr = 2\lambda$ Equation (16)

Solving equation 15 for Kc results in the following equations for critically damped tuning of a PID controller with Proportional and Integral action on an integrating process.

$Tr = 2\lambda$ Equation (17)

$Kc = \frac{4}{Kp*Tr}$ Equation (18)

Once again, using an example integrating process with a process gain of 0.001 %PV/Second/%OUT and with 0 dead time, the controller tuning for a chosen λ of 500 seconds would be as follows.

$$Tr = 2\lambda = 2 * 500\ Seconds = 1000\ seconds$$

$$Kc = \frac{4}{Kp * Tr} = \frac{4}{0.001\left(\frac{\frac{\%\mathrm{PV}}{\mathrm{Second}}}{\%\mathrm{OUT}} * (1000\ second)\right)} = 4.0\ \frac{\%OUT}{\%PV}$$

Figure 9 shows a simulation of the closed loop response to a step set point change for an integrating process with a process gain of 0.001%PV/Second/%OUT and 0 second deadtime combined with a Series or Standard PID controller that has a gain of 4.0 %OUT/%PV, 1000 seconds per repeat integral time and 0 derivative. Note that the PV initially reaches the new set point in the time λ after the step set point change. The PV then overshoots the SP by 13.5% and comes back to the SP in time 6* λ after the step SP change. This response is consistent regardless of the size of the step set point change if the output of the PID controller does not hit an upper or lower limit.

Figure 10 shows the simulated response to a step load disturbance of the outflow, Fo. As desired, the PV response is critically damped (no oscillations) and the offset between SP and

PV is reduced to 0 at steady state. Note that the calculated tuning parameters stop, or "arrest" the deviation of the PV from set point in the time λ after the step load disturbance. And, the PV is returned to the setpoint in time 6* λ after the step load disturbance without oscillations. This response is consistent regardless of the size of the step load disturbance change if the output of the PID controller does not hit an upper or lower limit.

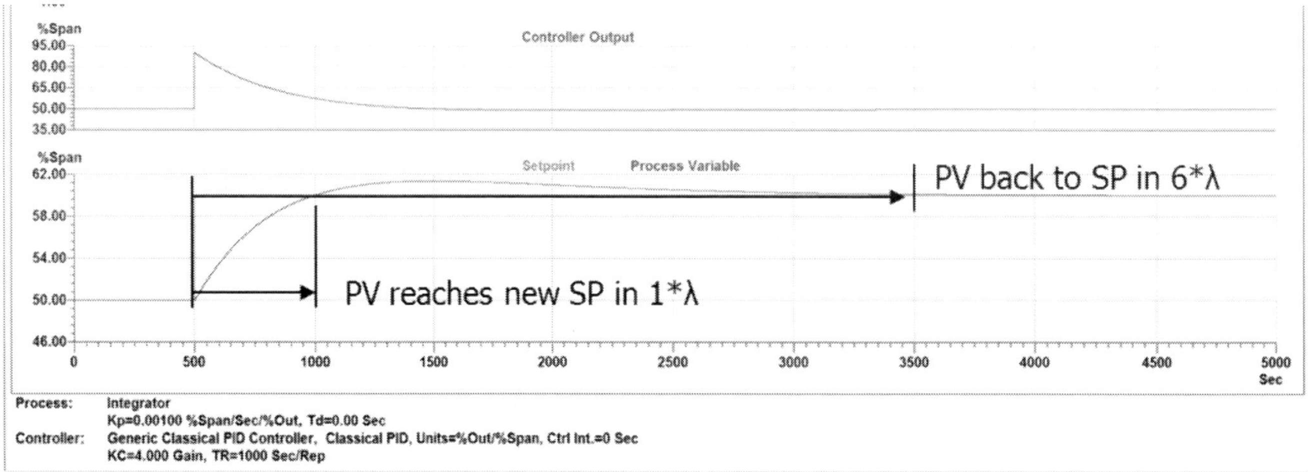

Figure 9 Simulation of Step Set Point Change for Integrating Process and PID Controller With Proportional and Integral Action, Tuned With a λ of 500 Seconds

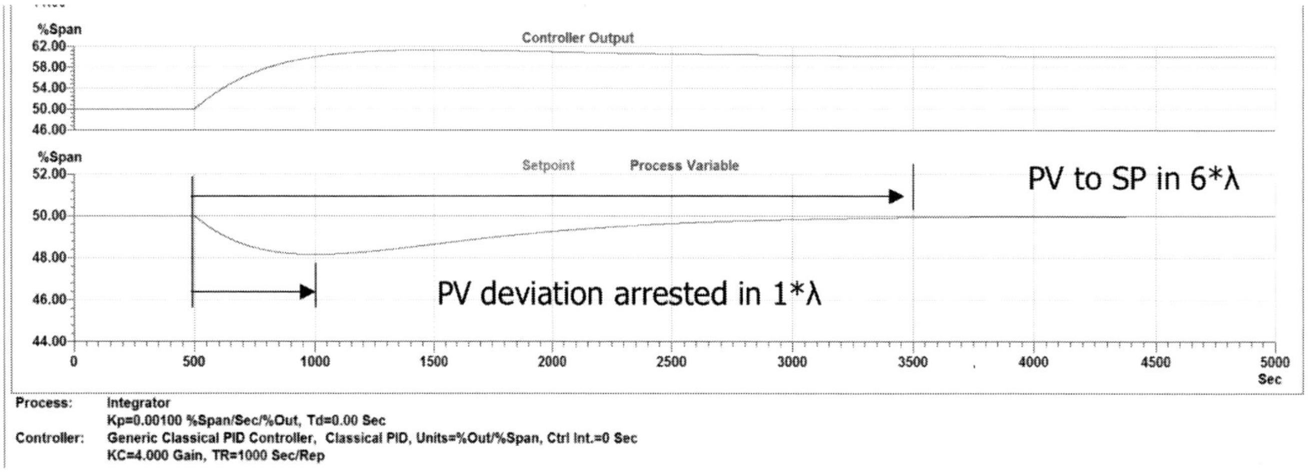

Figure 10 Simulation of Step Load Change for Integrating Process and PID Controller with Proportional and Integral Action Tuned With a λ of 500 Seconds

Case 3: Integrating Plus Deadtime Process with PI Controller

Now, consider the case of an integrating response with dead time, θ, as shown in Figure 3. Referring to Figure 5, the transfer function, Gp, of an integrating response with dead time, θ, is:

$$Gp(s) = \frac{Kp}{S} * e^{-s\theta} \qquad \text{Equation (19)}$$

The open loop transfer function of a series PID controller with proportional an integral action for the integrating plus dead time process shown in Figure 5 is:

$$Glp(s) = Gc(s) * Gp(s) = -\left[Kc * \left[1 + \frac{1}{TrS}\right]\right] * \left[\frac{Kp}{S} * e^{-S\theta}\right] = \frac{Kc(TrS+1)}{TrS} * Kp * \frac{e^{-S\theta}}{S} \qquad \text{Equation (20)}$$

$$Glp(s) = \frac{KcKp(TrS+1)e^{-S\theta}}{TrS^2} \qquad \text{Equation (21)}$$

The set point to PV transfer function is thus:

$$Gsp(s) = \frac{\frac{KcKp(TrS+1)e^{-S\theta}}{TrS^2}}{1+\frac{KcKp(TrS+1)e^{-S\theta}}{TrS^2}} \qquad \text{Equation (22)}$$

This can be simplified to the following:

$$Gsp(s) = \frac{(TrS+1)e^{-S\theta}}{\frac{Tr}{KcKP}*S^2+(TrS+1)*e^{-S\theta}} \qquad \text{Equation (23)}$$

Using the following first order Taylor expansion for dead time:

$$e^{-S\theta} \approx 1 - \theta S = -\theta S + 1 \qquad \text{Equation (24)}$$

Substituting this expansion, the set point transfer function equation 22 is simplified to:

$$Gsp(s) = \frac{\frac{KcKp(TrS+1)(-\theta S+1)}{TrS^2}}{1+\frac{KcKp(TrS+1)(-\theta S+1)}{TrS^2}} \qquad \text{Equation (25)}$$

And further simplified:

$$Gsp(s) = \frac{(TrS+1)(-\theta S+1)}{\left(\frac{Tr}{KcKp}-Tr\theta\right)S^2+(Tr-\theta)S+1} \qquad \text{Equation (26)}$$

Thus, the characteristic equation for the closed loop control system is:

$$Gsp(s) = \left(\frac{Tr}{KcKp} - Tr\theta\right)S^2 + (Tr - \theta)S + 1 \qquad \text{Equation (27)}$$

The solution to the quadratic equation (27) are the closed loop poles. To have a critically damped closed loop response, the controller tuning parameters Kc and Tr are chosen to have two closed loop poles at -1/λ. When this occurs, the resulting characteristic equation must be as shown in equation 14, repeated here:

$$0 = (\lambda S + 1)^2 = \lambda^2 S^2 + 2\lambda S + 1 \qquad \text{Equation (14)}$$

Comparing this to closed loop characteristic equation (27) results in the following equalities:

$$\frac{Tr}{KcKp} - Tr\theta = \lambda^2 \qquad \text{Equation (28)}$$

$$Tr - \theta = 2\lambda \qquad \text{Equation (29)}$$

Solving equation 28 and 29 for Kc and Tr, **results in the following equations for critically damped tuning of a Series or Standard PID controller with Proportional and Integral action on an integrating process with dead time.** Due to the approximation of dead time in the derivation of the tuning equations, lambda, λ, must be greater than the dead time.

Note that one cannot simply choose a small lambda, λ, for an integrating plus dead time and achieve the same closed loop performance as with an integrating process with 0 deadtime. The equations are valid for λ > θ, and preferably λ > 3*θ to provide reasonable margin of stability. In many cases, the lambda chosen for best process performance could be very large,

>>θ. An example of this is a level controller, on a tank whose capacity is used to absorb process variability, to reduce variability in the manipulated flow and downstream process.

$Tr = 2\lambda + \theta$	where λ>θ	Equation (30)
$Kc = \frac{Tr}{Kp*(\lambda+\theta)^2}$	where λ>θ	Equation (31)

Using an example integrating process with a process gain of 0.001 %PV/Second/%OUT and with 50 seconds dead time, the controller tuning for a chosen λ of 500 seconds would be as follows.

$$Tr = 2\lambda + \theta = (2 * 500 + 50)\, Seconds = 1050\ seconds$$

$$Kc = \frac{Tr}{Kp * (\lambda + \theta)^2} = \frac{1050\ seconds}{0.001\ \frac{\frac{\%\text{PV}}{\text{Second}}}{\%\text{OUT}} * \left((500 + 50) seconds\right)^2} = 3.47\ \frac{\%OUT}{\%PV}$$

Figure 11 shows a simulation of the closed loop response to a step set point change for an integrating process with a process gain of 0.001%PV/Second/%OUT and 50 second deadtime combined with a Series PID controller that has a gain of 3.47 %OUT/%PV, 1050 seconds per repeat integral time and 0 derivative. Note that the PV initially reaches the new set point in the time λ after the step set point change. The PV then overshoots the SP by 13.5% and comes back to the SP in time 6* λ after the step SP change. This response is consistent regardless of the size of the step set point change if the output of the PID controller does not hit an upper or lower limit.

Figure 12 shows the simulated response to a step load disturbance of the outflow, Fo. As desired, the PV response is critically damped (no oscillations) and the offset between SP and PV is reduced to 0 at steady state. Note that these tuning values stops, or "arrests" the deviation of the PV from set point in the time λ after the step load disturbance. And, the PV is returned to the setpoint in time 6* λ after the step load disturbance without oscillations. This response is consistent regardless of the size of the step load disturbance change if the output of the PID controller does not hit an upper or lower limit.

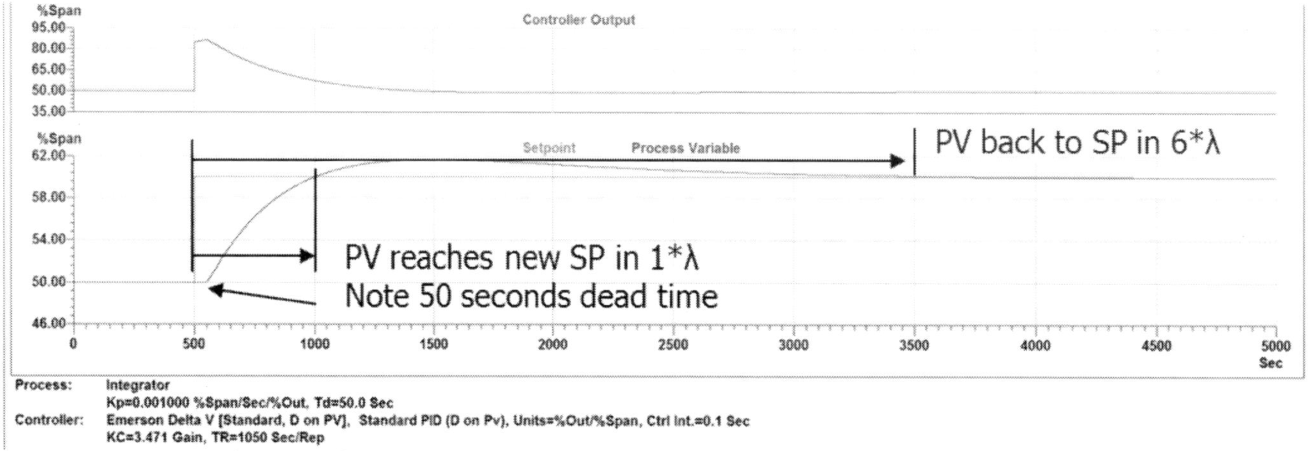

Figure 11 Simulation of Step Set Point Change for Integrating Process with Dead Time and PID Controller With Proportional and Integral Action, Tuned With a λ of 500 Seconds

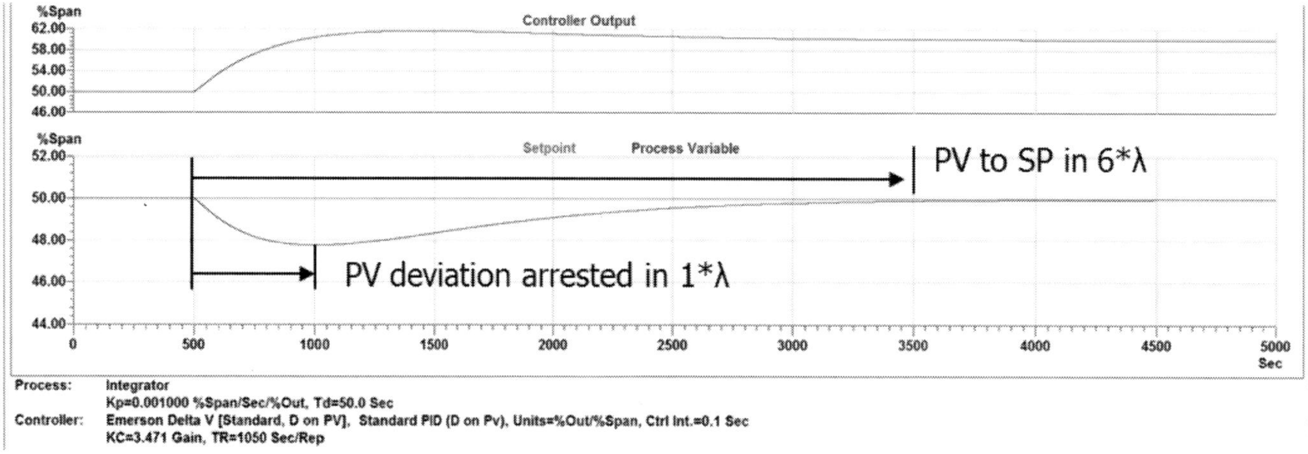

Figure 12 Simulation of Step Load Change for Integrating Process with Dead Time and PID Controller With Proportional and Integral Action Tuned With a λ of 500 Seconds

Case 4: Integrating Plus Deadtime and Lag Process with PID Controller

Now, consider the case of an integrating response with dead time, θ, plus lag, τ, as shown in Figure 3. Referring to Figure 5, the transfer function, Gp, of an integrating response with dead time, θ, plus lag, τ, is:

$$Gp(s) = \frac{Kp}{S} * \frac{1}{\tau S+1} * e^{-s\theta} \qquad \text{Equation (32)}$$

The transfer function, Gc, of a Series (aka Classical) PID controller with proportional, integral and derivative action is:

$$Gc(S) = Kc[TdS + 1] * \left[1 + \frac{1}{TrS}\right] \qquad \text{Equation (33)}$$

Where Kc has units of %OUT/%PV and Tr and Td have units of Seconds.

The open loop transfer function for the integrating process and series PID controller with proportional, integral action and derivative as shown in Figure 5 is:

$$Glp(s) = Gc(s) * Gp(s0 = [Kc[TdS + 1] * \left[1 + \frac{1}{TrS}\right]] * [\frac{Kp}{S} * \frac{1}{\tau S+1} * e^{-s\theta}] \qquad \text{Equation (34)}$$

Note that if the derivative time, Td, of the series PID controller is set equal to the process lag, τ, the two terms cancel each other and the open loop transfer function is:

$$Glp(s) = Kc * \left[1 + \frac{1}{TrS}\right] * \left[\frac{Kp}{S} * e^{-s\theta}\right] = \frac{KcKp*(TrS+1)*e^{-s\theta}}{TrS^2} \qquad \text{Equation (35)}$$

Note that equation (35) is identical to equation (21) which is the open loop transfer function for an integrating process with dead time and a PID controller with only proportional and integral action. Thus, by setting derivative time, Td, of the Series (aka Classical) PID controller equal to the process lag, τ, the controller gain, KC, and integral time, Tr, can be calculated in the same way as was done for an integrating process with dead time. Due to the approximation of dead time in the derivation of the tuning equations, lambda, λ, must be greater than the dead time.

Again, note that one cannot simply choose a small lambda, λ, and achieve the same closed loop performance as with a integrating process with 0 deadtime. The equations are valid for λ > θ, and preferably λ > 3*θ to provide reasonable margin of stability. In many cases, the lambda chosen for best process performance could be very large, >>θ.

Equations (36), (37) and (38) provide the tuning equations for a Series PID controller for an integrating process with dead time.

$Td = \tau$		Equation (36)
$Tr = 2\lambda + \theta$	where λ>θ	Equation (37)
$Kc = \frac{Tr}{Kp*(\lambda+\theta)^2}$	where λ>θ	Equation (38)

Using an example integrating process with a process gain of 0.001 %PV/Second/%OUT and with 50 seconds dead time, the controller tuning for a chosen λ of 500 seconds would be as follows.

$$Td = \tau = 50\ seconds$$

$$Tr = 2\lambda + \theta = (2 * 500 + 50)\ Seconds = 1050\ seconds$$

$$Kc = \frac{Tr}{Kp * (\lambda + \theta)^2} = \frac{1050\ seconds}{0.001\ \frac{\frac{\%\mathrm{PV}}{\mathrm{Second}}}{\%\mathrm{OUT}} * \left((500 + 50) seconds\right)^2} = 3.47\ \frac{\%OUT}{\%PV}$$

Figure 13 shows a simulation of the closed loop response to a step set point change for an integrating process with a process gain of 0.001%PV/Second/%OUT and 50 second deadtime combined with a Series PID controller that has a gain of 3.47 %OUT/%PV, 1050 seconds per repeat integral time and 50 seconds derivative time. Note that the PV initially reaches the new set point in the time λ after the step set point change. The PV then overshoots the SP by about 20% and comes back to the SP in time 6* λ after the step SP change. This response is consistent regardless of the size of the step set point change if the output of the PID controller does not hit an upper or lower limit.

Figure 14 shows the simulated response to a step load disturbance of the outflow, Fo. As desired, the PV response is critically damped (no oscillations) and the offset between SP and PV is reduced to 0 at steady state. Note that these tuning values stops, or "arrests" the deviation of the PV from set point in the time λ after the step load disturbance. And, the PV is returned to the setpoint in time 6* λ after the step load disturbance without oscillations. This response is consistent regardless of the size of the step load disturbance change if the output of the PID controller does not hit an upper or lower limit.

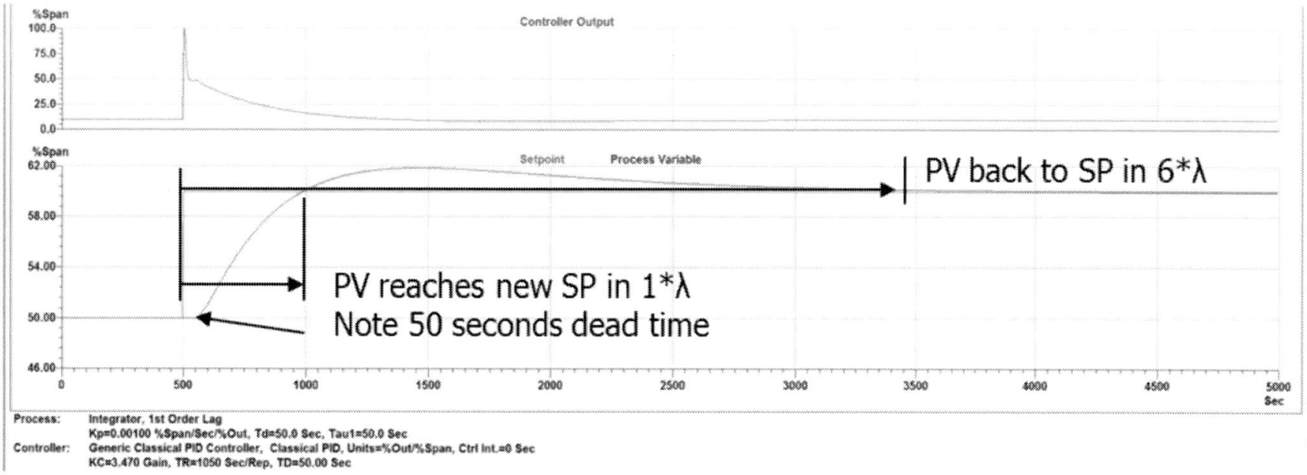

Figure 13 Simulation of Step Set Point Change for Integrating Process with Dead Time and Lag and a PID Controller With Proportional, Integral and Derivative Action, Tuned With a λ of 500 Seconds

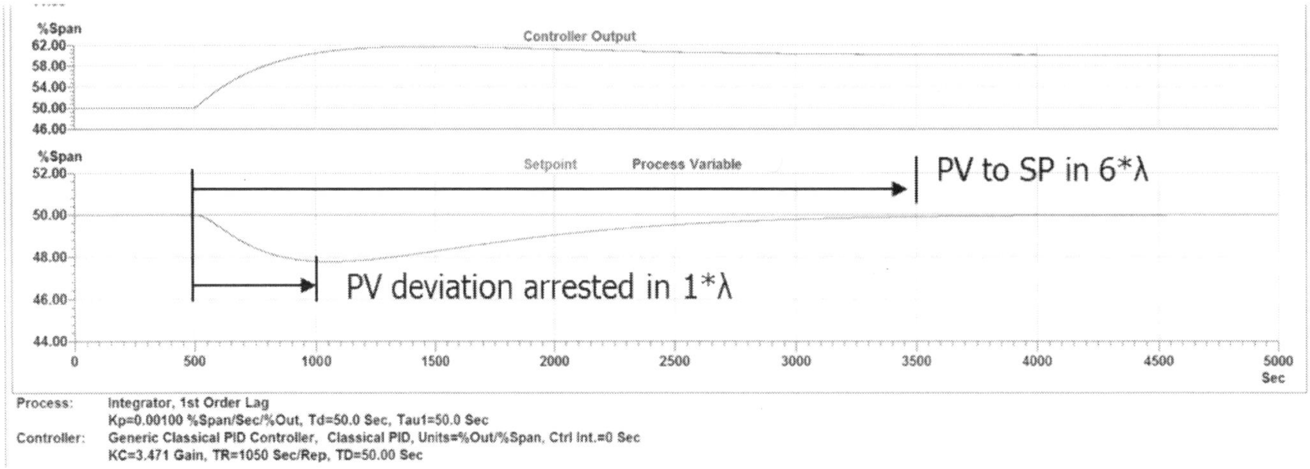

Figure 14 Simulation of Step Load Change for Integrating Process with Dead Time and Lag and a PID Controller With Proportional, Integral and Derivative Action, Tuned With a λ of 500 Seconds

Conversion of Tuning from Series to Standard PID

The tuning equations presented in this paper are based on a Series Implementation of the PID control algorithm. If the derivative time is set to 0 (no derivative action), the Series and Standard PID algorithms are identical. If the derivative time is not 0, the Series tuning parameters can be converted for the Standard PID algorithm.

$$Kc_{std} = Kc * \frac{Tr+Td}{Tr} \qquad \text{Equation (39)}$$

$$Tr_{std} = Tr + Td \qquad \text{Equation (40)}$$

$$Td_{std} = \frac{Tr*Td}{Tr+Td} \qquad \text{Equation (41)}$$

Summary

Tuning PID controllers for integrating process is often counter-intuitive. There are few tuning methods devoted to integrating process and most methods do not allow the choice of a closed loop speed of response. This feature is very useful when tuning the loop to maximize process performance vs. loop performance. This task is further complicated when the integrating response includes additional dynamics such dead time and lags. The tuning calculations presented in this paper are based in Internal Model Control theory where the tuning parameters are calculated to achieve the desired closed loop response. Note that tuning equations are based on a chosen transform for dead time which can impact the results but well within the range of variation of process models in the industrial process industry. The tuning equations have been used on thousands of control loops providing significant financial benefits.

There are additional cases of integrating process dynamics and tuning options which have not been explored in this paper. These cases can be discussed with the author if desired.

The tuning equations developed in this paper for a Series (aka Classical) PID Form are repeated here for convenience. There are many variations of implementation of the PID control function as well as many options. These variations and options may impact the suitability of these equations for a specific application. Good engineering practices should be used to verify the applicability of these equations for the application.

Process: Integrator

P only Controller: $Kc = \frac{1}{\lambda Kp}$ Equation (8)

PI Controller: $Tr = 2\lambda$ Equation (17)

(Series or Standard PID) $Kc = \frac{4}{Kp*Tr}$ Equation (18)

Process: Integrator plus dead time

PI Controller: $Tr = 2\lambda + \theta$ where λ>θ Equation (30)

(Series or Standard PID) $Kc = \frac{Tr}{Kp*(\lambda+\theta)^2}$ where λ>θ Equation (31)

Process: Integrator plus dead time and lag

Series PID Controller: $Td = \tau$ Equation (36)

$Tr = 2\lambda + \theta$ where λ>θ Equation (37)

$Kc = \frac{Tr}{Kp*(\lambda+\theta)^2}$ where λ>θ Equation (38)

512681 Tune Your PID Loops Please!

Abstract:

"Who looks after PID tuning at your site?" This is a question I often ask many professionals. The answers vary from a definitive deparment to "don't know" or "maintenance department", "instrumentation department" etc. The replies almost always suggest to me that there is not enough focus or attention paid to this important but generally forgotton layer of automation.

PID (Proportional-Integral-Derivative) controllers still form the backbone of automatic control systems in refineries, (petro) chemical installations, power plants and numerous other processes. The success of PID schemes lies in the simplicity of the implementation, reliability of the hardware and the stabilising effect on processes. It is however necessary to enter values for the "P", "I" and "D", referred to as tuning parameters, to make this PID controller function in a stable and efficient manner.

Significant benefits can be achieved by optimising the performance of PID loops, especially interacting loops. This presentation explores optimisation possibilities for process industries with some case studies on PID tuning and introduces a new paradign on how to undertake a PID tuning project at your facility.

511872 pH and Other Non-Linear Control Applications

Abstract:

The control of pH has many very important applications in industry. The non-linearity of pH control can make it extremely challenging using conventional PID control. This presentation examines non-linear control techniques to improve feedback pH control with applications to other non-linear control challenges. We will discuss techniques including several types of variable gains, linearization with ion concentration and titration curves. Change of variable will also be discussed. The advantages and disadvantages of each technique will be presented so that you can determine how to choose between the solutions to optimize performance in the intended application.

Cooperative Distributed Control: Model Predictive Control for the Industrial Internet of Things

AIChE 2018 Spring Meeting and 14th Global Congress on Process Safety
Session: Process Control and Optimization Developments - Wednesday, April 25, 2018

Dr Patrick Thorpe CEng FIChemE, patrick.thorpe@spirocontrol.com

Summary

Cooperative distributed model predictive control is receiving significant attention as a major next generation Model Predictive Control (MPC) technology and key component in unlocking the potential of the Industrial Internet of Things (IIoT). Cooperative distributed control enables a distributed network of cooperating controllers to collaborate towards a common objective without use of a centralized coordinator or optimizer. In addition to presenting a simple, modular route to plant-wide control and optimization, the technology offers a general approach to distributed MPC application within an IIoT framework.

Introduction

Model Predictive Control has been the de-facto advanced control technology used in the oil and gas sector to optimise refineries and petrochemical plants for several decades. Saving many millions of dollars per year in energy and increasing profits through improved plant throughput and product yields. However, due to its perceived high cost, complexity and restricted flexibility, MPC has seen limited adoption outside specific application areas. The inflexibility of existing MPC technology is particularly apparent when considering its application to large distributed but interdependent facilities. A further limiting factor has been the inability to incorporate future forecast of a changing environment into the control strategy. (Rawlings and Mayne 2013 provide a comprehensive overview of MPC theory).

Cooperative distributed control is an MPC based technology that overcomes some of these limitations. The solution enables a distributed network of controllers to collaborate towards a common objective without use of a centralized coordinator or optimizer, obeying all constraints of the network and utilizing available future planning forecasts such as production changes or environmental conditions. By eliminating the centralized component, the technology can work within complex and geographically distributed control structures. Crucially the technology lends itself to a plug and play architecture allowing a new sub-system model, variable, or hardware device to be added to and automatically assimilated into the cooperative network with relative ease. The embedded capability of cooperative control enables it to be used for applications that are not readily addressed by existing solutions including compressors, fired heaters, steam networks etc. In addition, the technology is designed to address a new range of control problems including gas distribution networks, building control, power grid load management etc. (See, for example, Stewart 2010, Razzanelli 2016, Shaoyuan 2015, Christofides 2013 for discussions of Cooperative Distributed MPC).

Outline of method

The proposed technology is designed to be integrated as a function block within an existing DCS controller, or on an edge device such as a Programmable Automation Controller (PAC) or Programmable Logic Controller (PLC). The code has been optimized for efficient execution and has a low memory footprint and has been developed in industry standard IEC 61131-3 allowing porting to many industrial controllers with minimal customization.

The unique architecture allows a multivariable control function to be distributed across many cooperating function blocks. This facilitates load distribution and ensures that peak CPU load is maintained within required limits. A large multivariable control solution can be deployed by interconnecting these simple function blocks. Cooperative controllers can also be distributed across multiple control processors.

What follows is a basic outline of the theory that facilitates cooperation between controllers. Consider a typical dynamic optimization function for a simple plant with two subsystems and two manipulated variables (u1 and u2):

$$V(x_1(0), x_2(0), u_1, u_2) = \sum_{i=1}^{N-1} \left[\frac{1}{2}[x_i, u_i]_i^T H_i [x_i, u_i]_i + f_i^T [x_i, u_i]_i\right] + [x_i, u_i]_N^T H_N [x_i, u_i]_N$$

$$x_{i+1} = Ax_i + B_1 u_i + B_2 u_2$$

$$P_i[x_i, u_i] \leq b_i$$

Each subsystem solves the above objective function for a known and fixed value of other subsystem input sequences. The process is repeated over a series of iterative steps.

1. (System –1) Fix u2 and solve with respect to u1

$$Set\ u_2 = u_2^p:\ \min_{u1} V(x_1(0), x_2(0), u_1, u_2) \rightarrow (u_1^*, u_2^p)$$

2. (System –2) Fix u1 and solve with respect to u2

$$Set\ u_1 = u_1^p:\ \min_{u2} V(x_1(0), x_2(0), u_1, u_2) \rightarrow (u_2^*, u_1^p)$$

3. Combine the results and make a step

$$u_i^{p+1} = \alpha(u_i^p) + (1-\alpha)(u_i^*)$$

The key design characteristic of cooperative control is that it solves the multivariable optimization without centralized coordination – each controller can operate autonomously and stably in the case of communication loss. Furthermore, cooperative iterations can terminate early (before full convergence) without compromising closed loop stability. With sufficient cooperative iterations the solution will converge to the global optimum and is equivalent to a single system wide multivariable controller.

There is significant flexibility in how the solution is deployed. Cooperative control *agents* can be separate multi-variable control applications configured on the same device or distributed over many *edge* devices. The edge devices may be geographically dispersed across a network.

Example 1: Plant-wide control and optimisation of an ethylene process

One of the use cases for the technology is to improve on current methods for controlling a large integrated process such as an ethylene plant or gas plant. Ethylene is commonly produced by steam cracking a petrochemical feedstock such as ethane. Figure-1 shows a simplified ethylene process schematic comprising a single steam cracking furnace, compressor, demethaniser column, expander, deethaniser column and C2-splitter column.

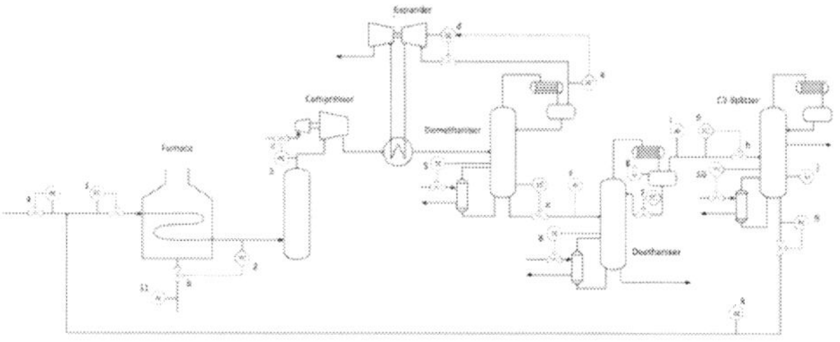

Fig 1. Simplified Ethylene Process Schematic

Each section of the ethylene process is considered a separate unit operation. It is commonplace for each unit operation to be controlled using a series of independent Proportional Integral Derivative (PID) control loops or model predictive controllers. However, due to the existence of strong interactions between unit operations in the steam cracking process, individual controllers are unable to achieve stable or optimal control. To overcome these challenges, the process may be controlled by a single monolithic model predictive controller or optimizer encompassing the complete process. Whilst this strategy can deliver significant benefits, there are substantial limitations relating to the complexity and engineering effort required to develop and maintain such an application.

In the proposed approach, separate control agents can be configured for each process section (steam cracking furnace, quench column, distillation columns etc.), thereby maintaining the simplicity of distributed control. However, the control agents are designed to operate cooperatively such that in addition to control of the local process section, each agent is also aware of global constraints and interactions between process sections. By operating cooperatively, control agents can approach a global optimum and ensure stability of each individual process section.

Example 2: Gas distribution network

Figure 2 shows a typical supply chain for the provision of ethane feed stock to a gas fed ethylene plant. Natural gas, which may be associated gas extracted with crude oil or non-associated gas, is fed to the gas plant for extraction of valuable products including ethane and higher hydrocarbons from the less valuable sales gas. NGL product is sent as a mixture of ethane, propane and butane by pipeline to the NGL plant where it is separated into its constituent components. The lightest component, ethane, is extracted in the first NGL column, which is the deethaniser, and sent as feed to the ethylene plant. It should be noted that each of the process plants (gas plant, NGL plant and ethylene plant) is a separate entity, with separate control facilities and process operators. Often the gas plant is geographically separated from the NGL plant with a pipeline supplying the NGL feed over large distances.

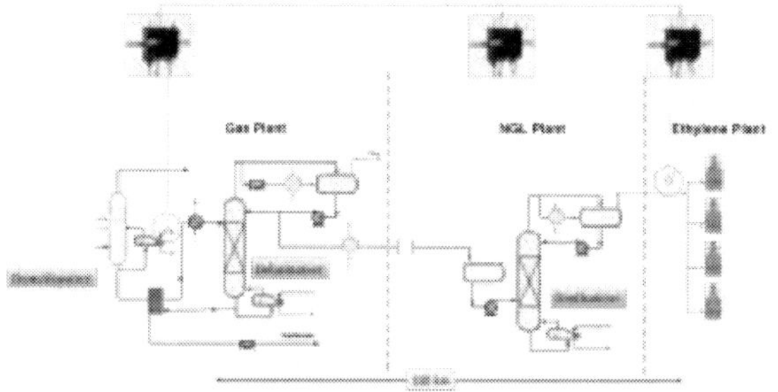

Fig 2. Ethylene Plant Feed Supply

Due to the geographic and operational separation, each plant is controlled independently with information shared only at the planning level. This approach ignores the dynamic interactions between the separate processes. For example, the control temperature of the gas plant demethaniser has a strong influence on the total ethane supply that is available to the ethylene plant. In the case of an operating disruption or feed reduction on the ethylene plant, the surplus ethane must be flared. In this situation it is far more economic to raise the gas plant demethaniser temperature, allowing the surplus ethane to be extracted with the local sales gas.

A cooperative distributed control solution covering the gas plant, NGL and ethylene plant with a shared objective would allow controllers on each plant to anticipate feed disruptions and minimize flaring while also maximizing the available ethane feed to the ethylene plant. This has the potential for significant financial benefits and a reduction in environmental emissions. The benefits in this case would be much higher than those derived from a steady state optimizer or planning model because of the need to respond quickly, in real-time, to shifting operating conditions.

Example 3: Utilities plant steam network

A typical process plant steam system may include a range of different steam generators connected to the network, all capable of producing steam at the required grade. However, different generators operate at different efficiencies. For example, there may be highly efficient Combined Heat and Power (CHP) units, less efficient gas fired boilers and finally oil-fired boilers with lower efficiency and higher emissions. The objective is to generate the required quantity of steam at the lowest total energy and within emissions limits.

It would be possible to optimize the overall steam network with a centralized optimizer that looked at the efficiency and generating capacity of each unit and coordinated the overall operation to maintain the required steam header pressure. But this centralized optimizer could be complex to develop and maintain and is rendered inoperable if new generators are added to the network.

With cooperative distributed control, each generator is treated as a separate sub-system with a local control agent. Each local control agent collaborates with others in the network so that collectively they agree on the best way to generate the required amount of steam. If there is a disruption to the steam network or one generator drops out, the remaining agents quickly re-coordinate to maintain the overall system. When a new steam generator is added it can be quickly assimilated into the network.

Example 4: Building control

Buildings consume more than 40% of the total primary energy resources throughout the world with an estimated 82% unrealized energy efficiency potential (IEA 2014). MPC is gaining popularity as an effective approach to building climate control given that it is a typical multi-variable constrained control problem. The main aim of MPC when applied to building management is to reduce the energy consumption, considering occupant comfort expectations.

Almost all buildings are controlled by independent zone controllers in a distributed architecture. However, this approach neglects the thermal coupling that exists between adjacent rooms, through internal walls, floors and door openings. It is for this reason that decentralized control schemes return a sub-optimal solution compared to centralized MPC. Moroşan et al. (2010) compared a conventional baseline controller with a decentralized and centralized MPC scheme, which achieved energy savings of 5.5% and 13.4%, respectively.

The limitation of centralized MPC is that as the system becomes more complex and spatially distributed, a significant amount of computing time and data transfer are required, leading to communication overhead. In case of communication breakdown, the controller is no longer able to compute the control signals. Consequently, for large buildings, the prevailing view is that despite the performance benefits of centralized control, for reliability issues it is still advisable to deploy a distributed architecture.

Cooperative zone controllers represent the best of both worlds accounting for thermal interactions and building wide constraints, enabling improved control with lower total energy. Additionally, cooperative controllers can incorporate common, building wide objectives such as a limit on the total power demand which may be particularly relevant when power is supplied by a renewable resource such as solar. The controllers can also take account of future planning data such as room occupancy scheduling, weather forecast and utility pricing tariffs. A simple, modular construction based on relatively simple, low cost hardware means cooperative distributed control presents significant advantages over a centralized building management system.

Example 5: Microgrids

A microgrid is a grouping of distributed energy sources with control capabilities that can work while connected to the main power network or can operate autonomously as dictated by demand. Microgrids range in size from large industrial or community systems, down to small localized systems.

Cooperative distributed control can be applied to perform optimal load management of a microgrid. Optimal load management can be achieved by implementing a moving time horizon window to plan the optimal grid strategy over a future time-period of up to 24-hours. Input data for calculating the optimal grid strategy can include forecast data on variables such as ambient temperature, solar intensity, wind speeds, expected load profile and electricity supply tariffs. In this way the system ensures the best use of available resources including energy storage devices whilst accounting for constraints such as total grid load and equipment operating limits. As the technology derives from closed-loop model predictive control, it means that the plan can be continuously updated in real-time based on a comparison of actual vs forecast data.

Figure 3 illustrates the power flows in a simple microgrid simulation comprising solar panels, conventional diesel generator, battery storage, load, and utility grid connection. The battery can store power when the grid price is low or when renewable output is high and can be made to discharge when the grid price is high or when the renewable output is low. Buying and selling with the utility grid is available based on the applicable tariff. There is an agreed cap on peak power consumed from the grid (12 kW). A conventional diesel generator is available to manage peak load. Each microgrid component is controlled by a separate cooperative control agent with agents cooperating to optimize the common objectives and constraints.

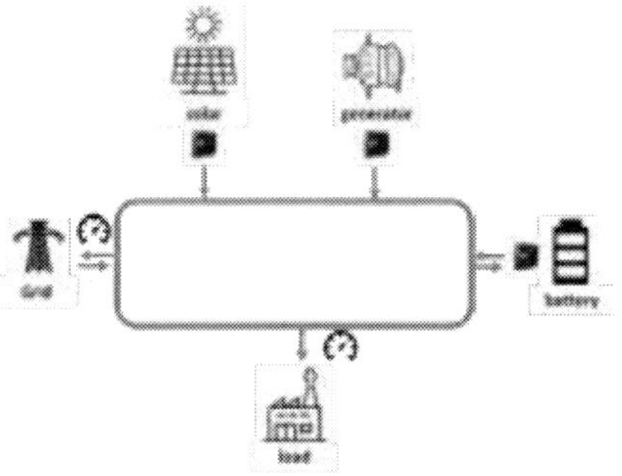

Fig 3. Simple microgrid illustration

Figure 4 shows typical forecast data for solar power generation, load profile and pricing tariff over a 24-hour period.

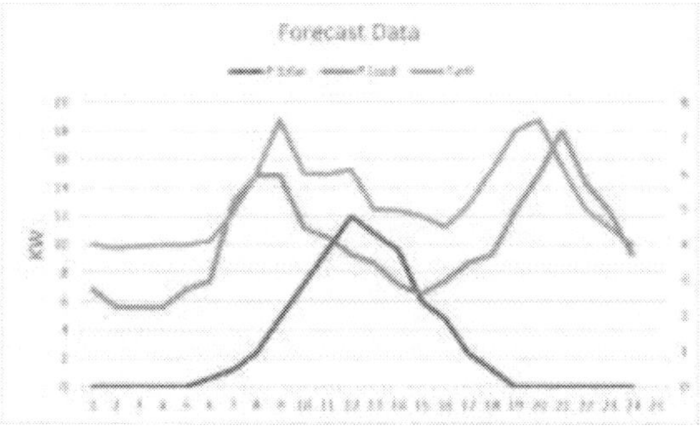

Fig 4. Forecast data

Cooperative controllers on the generator and battery can use the available forecast data, together with metered data throughout the day, to optimize overall power use. The benefit is shown in comparison to a reactive case where battery and generator use are determined by considering current load requirements, generated solar power, and spot tariff data without forecasting.

Figures 5 and 6 show the power drawn from the utility grid (positive values), power sold to the utility grid (negative values), battery state of charge (SOC) and the total normalized cost per operating hour. By forecasting the anticipated daily load and tariffs, the cooperative controllers make better use of the available battery storage capacity and conventional generator to lower the total electricity cost compared to the reactive case.

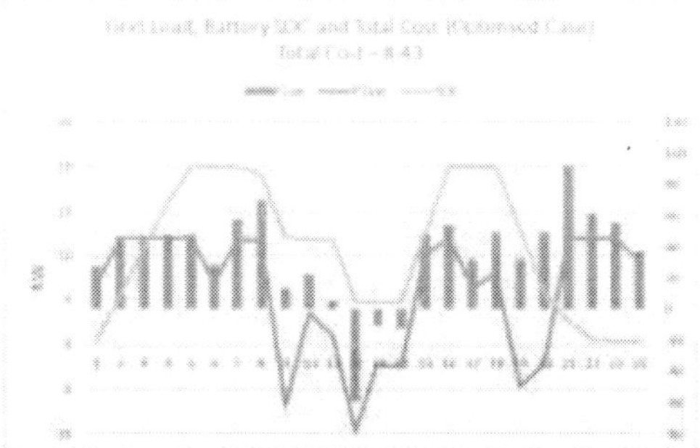

Fig 5. Optimized case

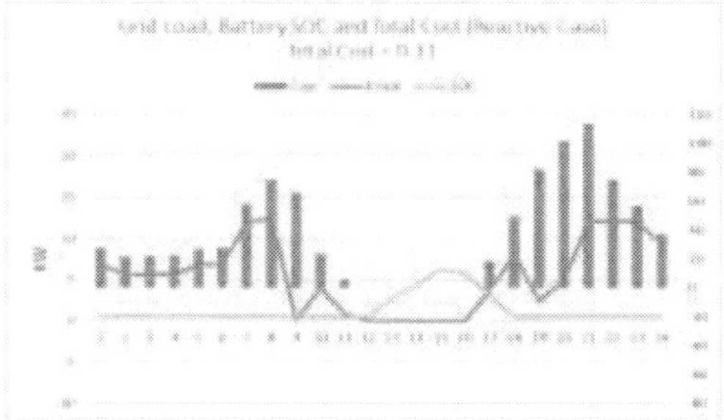

Fig 6. Reactive case

Conclusion

The Industrial Internet of Things (IIoT) offers unprecedented opportunities for industrial organizations to improve their operational efficiency. Hosted by IIoT enabled, open, secure, and interoperable system architectures, cooperative distributed control technology enables multiple interconnected and cooperating edge controllers to optimize an entire building, industrial process or utility network. This technology has the potential to harness the power of IIoT connectivity to transform process control and to deliver significant operational benefits. Additionally, it has the advantage of a simple, modular design and utilizes existing control standards and infrastructure.

References

Christofides, Scattolini, Muñoz de la Peña and Liu, Distributed model predictive control: A tutorial review and future research directions, Computers and Chemical Engineering, Volume 51, April 2013.

International Energy Agency (IEA) Capturing the Multiple Benefits of Energy Efficiency, 2014.

Moroşan, Bourdais, Dumur and Buisson. Building temperature regulation using a distributed model predictive control. Energy Build. 42 (2010) 1445-1452.

Razzanelli and Pannocchia, Parsimonious Cooperative Distributed MPC for Tracking Piece-Wise Constant Setpoints, IFAC Papers On Line, Volume 49, Issue 7, 2016, Pages 520-525

Rawlings and Mayne, Model Predictive Control: Theory and Design, Nob Hill Publishing, 2013.

Shaoyuan and Zheng, Distributed Model Predictive Control for Plant-Wide Systems, Wiley, 2015.

Stewart, Venkat, Rawlings, Wright and Pannocchia, Cooperative distributed model predictive control, Systems and Control Letters 59 (2010) 460-469.

513904 Automation of Knowledge Work in Advanced Process Control

Abstract:

Multivariable Predictive Control (MPC) has been an active area in both academic research and industrial applications for over 25 years. It still creates a tremendous value in the process industries by optimizing operations by increasing throughput, maximizing yields and reducing energy consumption. Although many advanced algorithms have been developed over the years in the area of model identification optimization and control, the methodology for the deployment of the MPC projects has not changed significantly and many of the MPC project steps are still manual and tedious, requiring significant effort and time. As the chemical processes and operation conditions continuously change, the process models need to be frequently re-identified and the controller tuning parameters need to be re-tuned in order to sustain the economic benefits of the MPC application.

Automation of knowledge work has been identified as one of the most impactful and disruptive initiatives with similar economic impact to the Industrial Internet of Things (IIOT). In this presentation we will discuss how automation of knowledge work applies to MPC in order to automate and simplify complex analysis and workflows and thus reduce barriers to deployment. More specifically we will look into automation of knowledge work for online Model Quality Analysis, Seed Model Generation, Automatic Data Slicing, Automatic Constrained Model Identification, Colinearity analysis and repair etc.

One particular area that we will focus on is automating the complex process of tuning the steady-state MPC optimizer via a Sequential Multi-Objective Optimization (SMO) that enables the user to simply specify the operational objectives of the process. An industrial example will be presented to demonstrate that via this approach the LP tuning remains consistent and does not need to be revisited when the controller models are updated.

513596 Inorganic Salts Removal from Hydraulic Fracturing Flowback Water By Cooling

Abstract:

This paper reported a novel method for removing inorganic salts from hydraulic fracturing flowback water by cooling. Both model compounds and real samples were tested. Model compounds study shows that the cooling effect for the following five salts are in the order of

$KCl > Sr(NO_3)_2 > NaCl > CsCl > CaCl_2$.

The real samples show that about 3% of inorganic salts were removed by cooling. This method can be applied to hydraulic fracturing flowback water recycling. The method could be improved the separation of oil from the water before the test.

512841 Metal-Hexacyanoferrate Based Nanoadsorbents for the Removal of Radioactive Cesium from Seawater

Abstract:

Although the nuclear power has been proposed as an alternative to thermal power due to the concern over global warming, it is also perceived as a threat due to the potential for the uncontrolled release of radionuclides. For example, the nuclear accident at Japan in 2011 released a huge quantity of radioactive contaminants into the environment. Among these, 137Cs is the most problematic contaminant due to its long half-life (30.2 years) and high-energy gamma ray emissions. Although various types of adsorbents such as zeolites, crystalline silicotitanate have been already developed, these adsorbents are either expensive or inefficient for large-scale remediation due to their low Cs selectivity. Alternatively, transition-metal hexacyanoferrates including Prussian blue, have been effectively used to remove 137Cs after Chernobyl nuclear accident due to their low cost and high selectivity for Cs. In recent years, various Prussian blue-functionalized magnetic nanoparticles have been reported for the efficient removal of Cs and magnetic recovery of adsorbent after use. However, Cs removal in seawater remains a large challenge due to high salt concentrations, and since the sorption capacity of PB for Cs is less than the sorption capacities of other metal (such as Cu, Ni, Co, and Zn)-hexacynoferrates.
To overcome these shortcomings, in the present study, various metal hexacyanoferrate-functionalized nanoadsorbents having enhanced colloidal stability in seawater were fabricated for the treatment of contaminated seawater. The resulting a few nanoadsorbents exhibited excellent adsorption capacity for Cs and also showed excellent removal efficiency of radioactive cesium (137Cs) from contaminated seawater.

512885 A Computational Model to Simulate Performance of Industrial Clarifiers

Abstract:

Sedimentation is one of the most popular water treatment processes and is used to separate solid particles from carrier fluid in settling tanks known as clarifiers. The clarifier as the last major facility in waste water treatment plant (WWTP) can affect and sometimes even limit the overall performance of the WWTP. An industrial-scale clarifier investigated in this work experienced lower than designed performance, which limited its wastewater handling capacity. The interested clarifier has a non-conventional feedwell design. The objective of this study was to analyze the ramifications of the current clarifier design and to come up with design modifications to improve the performance. A novel 3-dimensional unsteady CFD (computational fluid dynamics) model, where an effective and simplified approach was employed to address solids removal and slow-rotating rake (at about 1 rotation per hour), was developed to analyze the clarifier performance. The root causes for the unsatisfactory performance of the current clarifier design were identified. Subsequently, various design modifications were developed using the model to significantly improve the clarifier performance. This work shows that the present numerical model can become a computationally effective tool that will help researchers to design and optimize complex industrial clarifier geometries with rotating rakes in order to reduce the effluent TSS (total suspended solids) and optimize the clarifier performance.

513245 CFD Modeling of Waste Water Bio-Reactors

Abstract:

Computational Fluid Dynamic (CFD) modeling is a powerful tool for equipment design and development and is widely used at UOP, particularly for the multiphase systems. One such system is UOPâ€™s XCeed™ Bioreactor technology. Through CFD modeling, the hydrodynamic interaction of the air and water in the fixed-bed bioreactor was studied and the results were used to design a new large-scale wastewater treatment bioreactor. An innovative and efficient modelling method was developed to overcome a complex bioreactor configuration.

The XCeed™ bioreactor packing media has a proprietary solid phase which uniquely distributes high surface area within highly porous void spaces.

The critical challenge in this three-phase CFD model was finding an accurate approach to mimic wastewater (hydraulic) and aeration (gas-bubbles) patterns through the solid phase containing substantial biomass.

To better understand the flow pattern, flow distribution and the air bubble residence time inside the XCeed™ Bioreactor, the analysis was broken down into two sub-models.

Part 1 single-phase water-flow only analysis showed that the XCeed™ packing media caused the water flow to distribute uniformly across the packed-bed, even when high levels of biomass are present, without any channeling or by-pass or plugging.

Part 2 two-phase air/water flow analysis showed that the XCeed packing void areas affected immediate air flow expansion into and subsequent distribution throughout the high surface area of the packed bed, even with the accumulation of high biomass.

These CFD results confirm the effective benefits of UOPâ€™s distinctive approach to blending surface area with open space to enhance air/water distribution without plugging or channeling even when high biomass solids are present.

513356 Optimization of Water Demand Growth Fulfillment in Seawater Thermal Desalination System

Abstract:

In arid regions of the world such as Saudi Arabia, thermal desalination technologies such as Multiple Effect Distillation (MED) are the mainstream technologies for producing desalinated water for both the residential and industrial sectors. Thermal desalination technologies are generally characterized by its high capital intensity. Fixed cost charges in MED typically account for 40-50% of the unit cost of production compared to 30% in RO [1]. This poses a challenge in distressed thermal desalination network with an incessant water demand growth as the conventional approach of large grassroots expansion projects can lead to low capital productivity.

With the emergence of new technologies such as Membrane Distillation (MD) that can integrated with existing conventional thermal desalination units to boost water recovery, the problem of water demand growth fulfillment in a desalination network extends beyond searching for the optimal capacity expansions to finding the optimal portfolio of projects in the planning horizon. The considered technologies vary in their capital intensity, modularity, and performance calling for an overarching optimization framework of the capacity-planning problem.

This work develops the optimization framework for the fulfillment of water demand growth in a distressed desalination network considering a set of technological alternatives which are both discrete and continuous in nature. A superstructure representation and optimization formulation are introduced to simultaneously optimize the investment portfolio in the desalination network for the minimization of the total net present value (NPV), which includes both annualized fixed and operating costs as well as salvage values, over a specified planning horizon. The framework is applied on a case study for the fulfillment of a linear water demand growth and an existing MED desalination system. The results are reported using a Pareto chart to show changes in the optimal projects portfolio with key parameters changes.

AUTHOR INDEX

Abazajian, Armen 41
Abdeshova, Togzhan K. 2
Agarwal, Pratyush 139
Ahmadzadeh, Azita 175
Aldescu, Maria 65
Alghazal, Abdullah 126
Al-Jarba, Khalid 53
Al-Otaibe, Sultan 53
Alptekin, Gokhan 97
Atkinson, Brian 124
Baaqeel, Hassan 176
Baldea, Michael 57
Barney, Bryon 174
Beall IV, James 115, 143
Beebe, Dustin 3
Ben, Laurie 115, 125
Benson, Tracy J. 99
Bhargava, Manish 5, 58
Bishop, Ben 117
Botre, Chiranjivi 45
Burnak, Baris 52
Burton, Matthew 116
Butters, Tim 81
Carter, Elizabeth 121
Cernik, Randy J. 2
Chen, Lu 46, 96
Chianelli, Russell R. 99
Cima, Dennis L. 2
Cogswell, Kyle 129
Coletti, Francesco 36, 47, 88, 92
Cox, Pelin 140
Das, Shankhadeep 174
Dave, Sanjay 44
Desisto, William J. 128
Dhaliwal, Sam 164
Diaz-Bejarano, Emilio 36, 47, 88, 92
Eaton, Scott 128
Edgar, Thomas F. 57
El-Halwagi, Mahmoud 176
Erfan, Mani 98
Escarcega, John 13
Fazlollahi, Farhad 56
Feng, Maoqi 172
Fernandez, Sergio 122
Fleming, Brad 58
Garg, Ashutosh 51, 78
Gildert, Gary 98
Gill, Shabroz 166
Go, Kang Seok 141, 142
Gunukula, Sampath 128
Hall, Stephen 95, 126
Heniford, Richard R. 76
Ikarimoto, Cesar S. 12
Jansen, Christopher R. 123
Jeong, Sung Woo 141, 142
Jha, Meha 42, 64
Kalafatis, Alex 171
Kalita, Roomi 58
Kao, Jen-Hsiang 174
Karim, Nazmul 45
Kelly, Jeffrey D. 12, 34
Khogeer, Ahmed 118
Kidd, Mike 174
Kim, Kwang Ho 141, 142
Kim, Woo Hyun 141, 142
Kim, Yong Ku 142
Kim, Yong-Goo 141
Klein, Michael T. 139
Kuettel, Mark 174
Kumar, Prashant 98
Labanca, Leandro 96
Le Roux, Galo A. C. 12
Lee, Kune Woo 173
Lozano, Federico 92
Lynch, David 98
Macchietto, Sandro 36, 47, 88, 92
Mangan, Craig 13
Mansouri, Majdi 45
Manthanwar, Amit 52
McBride, Will 3
Menezes, Brenno C. 12, 34
Mergenbayev, Askhat M. 2
Mistry, Shashi 13
Mohedas, Sergio 79
Mostofi, Reza 175
Nho, Nam Sun 141, 142
Nolan, Tom 165
Nordh, Perry 13
Nounou, Hazem 45
Nounou, Mohamed 45
Ochoa-Estopier, Lluvia M. 96
Ogumerem, Gerald S. 52
Olsen, Tim 35, 80
Onome, Akpomiemie Mary 46
Osta, Sebastien 11
Ostraat, Michele L. 122
Park, Chan Woo 173
Pendse, Hemant P. 128
Peters, Nate 13
Pistikopoulos, Efstratios N. 52
Poklop, Steve 175
Prashant, Kumar 95
Raftery, Jonathan P. 45
Rallapalli, Jagan Mohan 54
Ramkumar, Shwetha 55
Rao, Samvith 77
Rasmussen, Kent H. 2
Ratan, Sanjiv 132
Reis, Lucas 171
Ronces, Edgar Amaro 120

AUTHOR INDEX

Saudagar, Munawar 53, 79
Schroeder, Scott 43, 119, 123
Shah, Ashish 5
Shatto, Don P...55
Shelton, Al 65
Shire, Tim 4
Silva, Kristin...172
Sloley, Andrew W..................................40, 41, 119, 123
Smith, Robin..46
Sunol, Aydin K. 129
Thackston III, Bill...117
Thorpe, Patrick 166
Trenchard, Andrew 44
Valentine, Julie..42, 64
Van Walsum, G. Peter 128
Varraveto, Dominic..127
Wang, Dian 13
Wang, Sujing 1
Webb, Christopher 44
Wheeler, M. Clayton 128
William, Whyman 131
Wold, Kjell...35
Xu, Jialin..1
Xu, Kexin..46
Xu, Qiang...1
Yan, Lingqing..57
Yang, Hee-Man 173
Ye, Minghua 53, 79
Yeh, Norman K. 55
Zhang, Jian 1
Zhang, Ke 122
Zhao, Hong 171
Zink, Steven F. 130